Student Solutions Manual for Tussy and Gustafson's *Prealgebra*

Second Edition

Tod Shockey

Kimberly Shockey

BROOKS/COLE
THOMSON LEARNING™

Australia • Canada • Mexico • Singapore • Spain • United Kingdom • United States

BROOKS/COLE
THOMSON LEARNING™

Assistant Editor: *Rachael Sturgeon*
Marketing Manager: *Leah Thomson*
Marketing Communications: *Samantha Cabaluna*
Marketing Assistant: *Maria Salinas*
Editorial Assistant: *Lisa Jones*
Production Coordinator: *Dorothy Bell*
Print Buyer: *Christopher Burnham*
Printing and Binding: *Globus Printing*

For more information about this or any other Brooks/Cole product, contact:
BROOKS/COLE
511 Forest Lodge Road
Pacific Grove, CA 93950 USA
www.brookscole.com
1-800-423-0563 (Thomson Learning Academic Resource Center)

Printed in the United States of America

10 9 8 7 6 5 4 3 2 1

ISBN 0-534-38328-9

TABLE OF CONTENTS

Section 1.2 Adding and Subtracting Whole Numbers

Vocabulary

1. When two numbers are added, the result is called a **sum**. The numbers that are added are called **addends**.
3. A **rectangle** is a four-sided figure (like a dollar bill) whose opposite sides are of equal length.
5. When two numbers are subtracted, the result is called a **difference**. In a subtraction problem, the **subtrahend** is subtracted from the **minuend**.
7. The property that allows us to group numbers in an addition in any way we want is called the **associative** property of addition.

Concepts

9. Commutative property of addition
11. Associative property of addition
13. Use x and y,
    a. Commutative property of addition: $x+y=y+x$
    b. Associative property of addition: $(x+y)+z=x+(y+z)$

15. Any number added to **0** remains the same.
17. The addition fact illustrated is $4+3=7$.

Notation

19. The grouping symbols ( ) are called **parentheses**.

21. $(36+11)+5=47+5$
    $=52$

Practice

23. $25+13=38$

25. $156+305=461$

27. $19+39+53=111$

29. $(95+16)+39=(111)+39=150$

31. $25+(321+17)=25+(338)=363$

33. $\begin{array}{r} 632 \\ +347 \\ \hline 979 \end{array}$

65. COFFEE

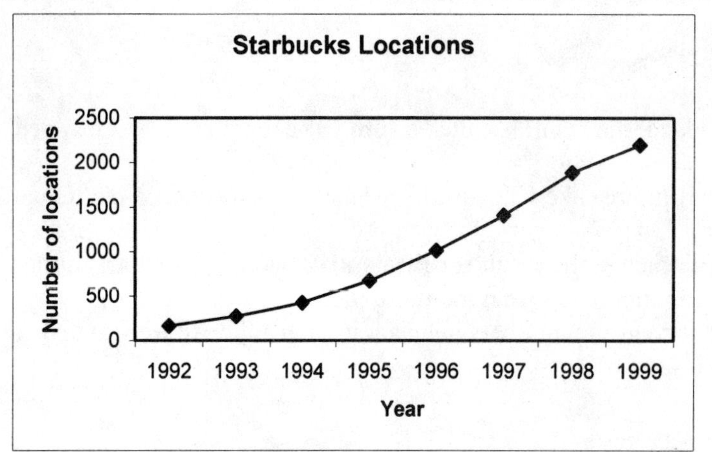

67. CHECKS

c. $\$15,601\frac{00}{100}$; Fifteen thousand six hundred one and $\frac{00}{100}$.

d. $\$3,433\frac{46}{100}$; Three thousand four hundred thirty three and $\frac{46}{100}$.

69. EDITING

one million, eight hundred sixty-five thousand, five hundred ninety-three votes; 1,865,593
four hundred eighty-two thousand, eight hundred eighty; 482,880
one thousand five hundred three; 1,503
two hundred sixty-nine; 269
forty-three thousand four hundred forty-nine;43,449

71. SPEED OF LIGHT

a. Rounded to the nearest hundred thousand meters per second is 299,800,000 m/s.
b. Rounded to the nearest million meters per second is 300,000,000 m/s.

Writing

73. Answers will vary.
75. Answers will vary.

51. 5,926,000
53. 5,900,000
55. $419,160
57. $419,000

Applications

59. EATING HABITS

Country	*Per Person Consumption*
United States	261 lb
New Zealand	259 lb
Australia	239 lb
Cyprus	236 lb
Uruguay	230 lb
Austria	229 lb
Saint Lucia	222 lb
Denmark	219 lb
Canada	211 lb
Spain	211 lb

61. MISSIONS TO MARS
 a. The 70's had the greatest number of successful missions.
 b. The 60's had the greatest number of unsuccessful missions.

63. ENERGY RESERVES

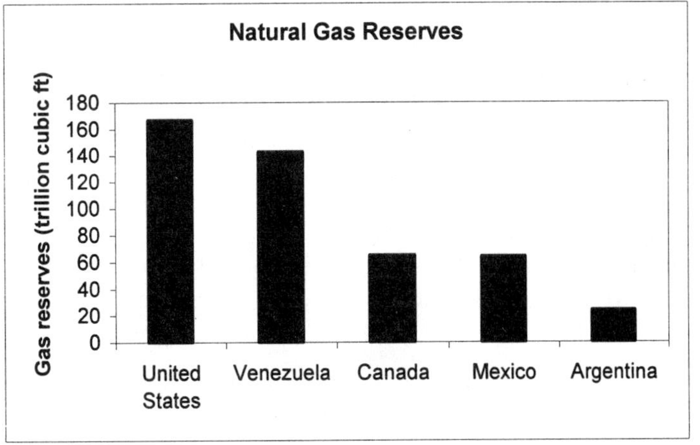

Section 1.1 An Introduction to the Whole Numbers

Vocabulary

1. A **set** is a collection of objects.
3. When 297 is written as 2 hundreds + 9 tens + 7 ones, it is written in **expanded** notation.
5. Using a process know as graphing, whole numbers can be represented on a **number** line.

Concepts

7. In the numeral 57,634, the **3** is in the tens column.
9. In the numeral 57,634, the **6** is in the hundreds column.
11. whole numbers
13. Graph 1, 3, 5, and 7

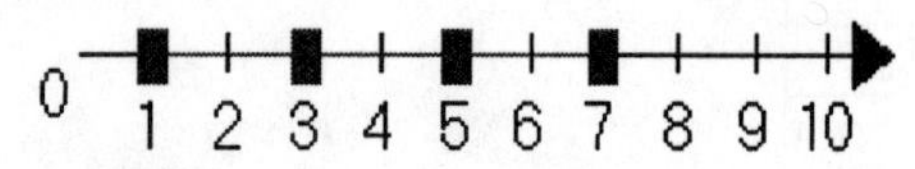

15. Graph the whole numbers less than 6

17. $47 > 41$
19. $309 > 300$
21. $2052 < 2502$
23. Since $4 < 7$, it is also true that $7 > 4$.

Notation

25. The symbols { }, called **braces**, are used when writing a set.

Practice

27. 245; 2 hundreds + 4 tens + 5 ones; two hundred forty-five
29. 3609; 3 thousands + 6 hundreds + 9 ones; three thousand six hundred nine
31. 32,500; 3 ten thousands + 2 thousands + 5 hundreds; thirty-two thousand five hundred
33. 104,401; 1 hundred thousand + 4 thousands + 4 hundreds + 1 one;
    one hundred four thousand four hundred one
35. 425
37. 2736
39. 456
41. 27,598
43. 9,113
45. 10,700,506
47. 79,500
49. 80,000

35. $\begin{array}{r} 1{,}372 \\ +\ 613 \\ \hline 1{,}985 \end{array}$

37. $\begin{array}{r} 6{,}427 \\ +3{,}573 \\ \hline 10{,}000 \end{array}$

39. $\begin{array}{r} 8{,}539 \\ +7{,}368 \\ \hline 15{,}907 \end{array}$

41. $\begin{array}{r} 1{,}246 \\ 578 \\ +\ 37 \\ \hline 1{,}861 \end{array}$

43. $\begin{array}{r} 3{,}156 \\ 1{,}578 \\ +\ 578 \\ \hline 5{,}312 \end{array}$

45. $\text{Perimeter} = 2l + 2w$

$2(32) + 2(12) = 88 \text{ ft}$

47. $\text{Perimeter} = 4s$

$4(17) = 68 \text{ in}$

49. $17 - 14 = 3$

51. $39 - 14 = 25$

53. $174 - 71 = 103$

55. $633 - (598 - 30) = 633 - (568) = 65$

57. $160 - 15 - 4 = 160 - 19 = 141$

59. $29 - 17 - 12 = 29 - 29 = 0$

61. $\begin{array}{r} 367 \\ -343 \\ \hline 24 \end{array}$

63. $\begin{array}{r} 423 \\ -305 \\ \hline 118 \end{array}$

65. $\begin{array}{r} 1,537 \\ -\ \ 579 \\ \hline 958 \end{array}$

67. $\begin{array}{r} 4,267 \\ -2,578 \\ \hline 1,689 \end{array}$

69. $\begin{array}{r} 17,246 \\ -\ \ 6,789 \\ \hline 10,457 \end{array}$

71. $\begin{array}{r} 15,700 \\ -15,397 \\ \hline 303 \end{array}$

73. $43-12+9=43-3=40$

75. $120+30-40=120-10=110$

Applications

77. TAXIS
$\$23-\$5=\$18$
$18 was the fare.

79. MAGAZINE CIRCULATION
Circulation in 1996 was 1,803,566
Circulation in 1997 grew by 15,865
Circulation in 1998 decreased by 69,404

Total circulation in 1997 was $1,803,566+15,865=1,819,431$
Total circulation in 1998 was $1,819,431-69,404=1,750,027$

The total circulation in 1998 was 1,750,027.

81. BANKING
Initial balance was $370.
Deposit of $40
Withdrawal of $197

Therefore, after the deposit and withdrawal the account balance is $\$370+\$40-\$197=\$410-\$197=\$213$.

83. TAX DEDUCTION

The total number of miles driven is the sum of the miles driven for each of the first six months which is

$2,345+1,712+1,778+445+1,003+2,774$

$=2,345+1,712+1,778+445+3,777$

$=2,345+1,712+1,778+4,222$

$=2,345+1,712+6,000$

$=2,345+7,712$

$=10,057$ miles

85. INCOME

c. Step 1/Column 1

Each year is a Step so sum the salaries for the first five steps from Column 1, $\$26,785+\$28,107+\$29,429+\$30,751+\$32,073=\$147,145$.

d. Step 1/Column 3

Each year is a Step so sum the salaries for the first five steps from Column 3, $\$29,701+\$31,023+\$32,345+\$33,667+\$34,989=\$161,725$.

87. BLUEPRINT

The length of the house is the sum of the parts, $24+35+16+16=91$ ft.

The house is 91 ft long

89. CAR EMISSIONS

The number of tons that have been removed daily is the sum of the tons removed daily for each step taken. $215+85+117+35+70+150+120=792$ tons

792 tons have been removed daily because of the legislation.

91. CITY FLAG

The number of inches of fringe needed is equivalent to the perimeter of the flag.

$P=2l+2w=2(64)+2(34)=128+68=196$ in.

The amount of fringe needed is 196 inches.

Writing

93. Answers will vary.

Review

95. 3,125; 3 thousands + 1 hundred + 2 tens + 5 ones

97. Rounded to the nearest ten is 6,354,780.

99. Rounded to the nearest ten thousand is 6,350,000

Section 1.3 Multiplying and Dividing Whole Numbers

Vocabulary

1. **Multiplication** is repeated addition.
3. The statement $ab = ba$ expresses the **commutative** property of multiplication.
5. If a square measures one inch on each side, its area is 1 **square inch**.

Concepts

7. $8+8+8+8=4\cdot 8$

9. The amount of surface enclosed by a rectangle is found by multiplying its length by its width.

11. Do each multiplication
    a. $1\cdot 25=25$
    b. $62(1)=62$
    c. $10\cdot 0=0$
    d. $0(4)=0$

13. The number of squares is $5\cdot 12$.

Notation

15. Three symbols used for
    a. Multiplication are $\times$, $\bullet$, $(\ )$.
    b. Division are $\div$, $-$, $\overline{)\ \ }$.

17. ft^2 means square feet.

Practice

19. $12\cdot 7=84$

21. $27(12)=324$

23. $9\cdot(4\cdot 5)=9\cdot 20=180$

25. $5\cdot 7\cdot 3=35\cdot 3=105$

27.
$$\begin{array}{r} 99 \\ \times \underline{77} \\ 693 \\ \underline{693} \\ 7,623 \end{array}$$

29.
$$\begin{array}{r} 20 \\ \times \underline{53} \\ 60 \\ \underline{100} \\ 1,060 \end{array}$$

31.
$$\begin{array}{r} 112 \\ \times \underline{23} \\ 336 \\ \underline{224} \\ 2,576 \end{array}$$

33.
$$\begin{array}{r} 207 \\ \times \underline{97} \\ 1449 \\ \underline{1863} \\ 20,079 \end{array}$$

35. $13,456 \cdot 217 = 2,919,952$

37. $3,302 \cdot 358 = 1,182,116$

39.
$$\begin{aligned} Area &= lw \\ &= 14(6) \\ &= 84\,\text{in}^2 \end{aligned}$$

41.
$$\begin{aligned} Area &= lw \\ &= 12(12) \\ &= 144\,\text{in}^2 \end{aligned}$$

43. $40 \div 5 = 8$

45. $42 \div 14 = 3$

47. $132 \div 11 = 12$

49. $\frac{221}{17} = 13$

51. $\begin{array}{r} 73 \\ 13\overline{)949} \\ \underline{91} \\ 39 \\ \underline{39} \\ 0 \end{array}$

Answer is 73.

53. $\begin{array}{r} 41 \\ 33\overline{)1,353} \\ \underline{132} \\ 33 \\ \underline{33} \\ 0 \end{array}$

Answer is 41.

55. $\begin{array}{r} 205 \\ 39\overline{)7,995} \\ \underline{78} \\ 19 \\ \underline{0} \\ 195 \\ \underline{195} \\ 0 \end{array}$

Answer is 205.

57. $\begin{array}{r} 210 \\ 29\overline{)6,090} \\ \underline{58} \\ 29 \\ \underline{29} \\ 00 \\ \underline{0} \\ 0 \end{array}$

Answer is 210.

59. $\begin{array}{r} 8 \\ 31\overline{)273} \\ \underline{248} \\ 25 \end{array}$

The quotient is 8 and the remainder is 25.

61. $\begin{array}{r} 20 \\ 37\overline{)743} \\ \underline{74} \\ 3 \\ \underline{0} \\ 3 \end{array}$

The quotient is 20 and the remainder is 3.

63. $\begin{array}{r} 30 \\ 42\overline{)1,273} \\ \underline{126} \\ 13 \\ \underline{0} \\ 13 \end{array}$

The quotient is 30 and the remainder is 13.

65. $\begin{array}{r} 31 \\ 57\overline{)1,795} \\ \underline{171} \\ 85 \\ \underline{57} \\ 28 \end{array}$

The quotient is 31 and the remainder is 28.

Applications

67. FIGURING WAGES
The cook worked 12 hours at $11 per hour, therefore she earned $12(11) = \$132$.

69. FINDING DISTANCE
The car can hold 14 gallons and goes 29 miles on one gallon; therefore on a full tank (14 gallons) the car can go $14(29) = 406$ miles.

71. CONCERT ATTENDANCE

The total number of concerts is $2(37) = 74$. Since 1700 fans attended each concert the total number of people to hear the group was $1700(74) = 125,800$.

73. ORANGES IN JUICE

If it takes 13 oranges to make one can of juice then $24(13) = 312$ oranges are need to make a case of 24 cans.

75. CAPACITY OF AN ELEVATOR

If there are 14 people each averaging 150 pounds then the weight in the elevator is $14(150) = 2100$ pounds. The elevator is overloaded.

77. WORD PROCESSING

If there are 8 columns and 9 rows then there will be $8(9) = 72$ entries in the table.

79. DISTRIBUTING MILK

$$\begin{array}{r} 3 \\ 23\overline{)73} \\ \underline{69} \\ 4 \end{array}$$

If the milk is distributed evenly then there will be **4 cartons** left over.

81. MILEAGE

$$\begin{array}{r} 5 \\ 140\overline{)700} \\ \underline{700} \\ 0 \end{array}$$

The bus can travel **5 miles** on one gallon.

83. CONVERSION

$$\begin{array}{r} 2 \\ 5,280\overline{)11,000} \\ \underline{10,560} \\ 440 \end{array}$$

11,000 is **440 feet** more than two miles.

85. PRICE OF A TEXTBOOK

$954,193 \div 23,273 = 41$

The cost of each book is $41.

87. VOLLEYBALL LEAGUE

Since a reasonable team size is 7 to 10 girls, divide the total number of girls by each team size to determine which meets all the requirements listed.

$216 \div 7 = 30$ teams with a remainder of 6 girls

$216 \div 8 = 27$ teams

$216 \div 9 = 24$ teams

$216 \div 10 = 10$ teams with a remainder of 6 girls

Since all teams must have the same number of players and the number of teams must be even, there should be **24 teams with 9 girls on each** to meet all the requirements.

89. COMPARING ROOMS

$$Area_{\text{rectangle}} = lw \qquad Area_{\text{square}} = lw$$
$$= 14(17) \qquad = 16(16)$$
$$= 238 \text{ feet}^2 \qquad = 256 \text{ feet}^2$$

The **square room** has the greater area.

$$Perimeter_{\text{rectangle}} = 2l + 2w \qquad Perimeter_{\text{square}} = 4l$$
$$= 2(14) + 2(17) \qquad = 4(16)$$
$$= 28 + 34 \qquad = 64 \text{ feet}$$
$$= 62 \text{ feet}$$

The **square room** has the greater perimeter.

91. GARDENING

$$Area_{garden} = lw$$
$$= 27(19)$$
$$= 513 \text{ ft}^2$$

Since the path is 125 ft^2 there will be $513 \text{ ft}^2 - 125 \text{ ft}^2 = 388 \text{ ft}^2$ left for planting.

Writing

93. Answers will vary.
95. Answers will vary.

Review

97. The 8 is in the hundreds column.

99.
$$\begin{array}{r} 357 \\ 39 \\ +476 \\ \hline 872 \end{array}$$

Estimation

1. $$\begin{array}{r} 30,000 \\ 10,000 \\ 9,000 \\ 1,000 \\ 10,000 \\ +\ 30,000 \\ \hline 90,000 \end{array}$$
   The estimate does not seem reasonable; no.

3. $$\begin{array}{r} 500 \\ \times\ \ 70 \\ \hline 35,000 \end{array}$$
   The estimate does not seem reasonable; no.

5. $60,000 \div 30 = 2,000$
   The answer given of 200 is not reasonable.

7. CAMPAIGNING
   She flew approximately 8,900 miles.
   $4,000 + 600 + 1,000 + 300 + 3,000 = 8,900$ mi

9. GOLF COURSE
   The number of bags needed is approximately $90,000 \div 3,000 = 30$.

11. CURRENCY
   The number of \$5 bills in circulation is approximately $8,000,000,000 \div 5 = 1,600,000,000$.

Section 1.4 Prime Factors and Exponents

Vocabulary

1. Numbers that are multiplied together are called **factors**.
3. To **factor** a whole number means to express it as the product of other whole numbers.
5. Whole numbers, greater than 1, that are not prime numbers are called **composite** numbers.
7. To prime factor a number means to write it as a product of only **prime** numbers.
9. In the exponential expression 6^4, 6 is called the **base**, and 4 is called the **exponent**.

Concepts

11. $1 \cdot 27$ and $3 \cdot 9$

13. Find the number
    a. 44
    b. 100

15. Find the factors
    a. 1 and 11
    b. 1 and 23
    c. 1 and 37
    d. All the numbers are prime.

17. If 4 is a factor of a whole number then 4 also divides that number exactly.

19. $2 \cdot 3 \cdot 3 \cdot 5 = 90$

21. $11^2 \cdot 5 = 11 \cdot 11 \cdot 5 = 605$

23. The order of the base and exponent in an exponential expression **cannot** be changed.

    $3^2 \neq 2^3$ since $3^2 = 9$ and $2^3 = 8$

25. Prime factorization of 30 is $2 \cdot 3 \cdot 5$

```
    30
   ↙ ↘
  2   15
     ↙ ↘
    3   5
```

Prime factorization of 242 is $2 \cdot 11 \cdot 11$

```
    242
   ↙ ↘
  2   121
     ↙ ↘
    11  11
```

The prime factor in common for 30 and 242 is 2.

27. Prime factorization of 20 is $2 \cdot 2 \cdot 5$

20
↙↘
2 10
  ↙↘
  2 5

Prime factorization of 50 is $2 \cdot 5^2$

50
↙↘
2 25
  ↙↘
  5 5

The prime factors in common for 20 and 50 are 2 and 5.

29.

150
↙↘
30 5
↙↘
2 15
  ↙↘
  3 5

150
↙↘
15 10
↙↘ ↙↘
3 5 2 5

The results are the same.

31. Complete the table

Product of factors of 12	Sum of factors of 12
$1 \cdot 12$	$1+12=13$
$2 \cdot 6$	$2+6=8$
$3 \cdot 4$	$3+4=7$

33. If the number is even the first choice when doing a prime factorization should be 2.

Notation

35. $7^3 = 7 \cdot 7 \cdot 7$
37. $3^5 = 3 \cdot 3 \cdot 3 \cdot 3 \cdot 3$
39. $5^2(11) = 5 \cdot 5 \cdot 11$
41. $10^1 = 10$
43. $2 \cdot 2 \cdot 2 \cdot 2 \cdot 2 = 2^5$
45. $5 \cdot 5 \cdot 5 \cdot 5 = 5^4$
47. $4(4)(5)(5) = 4^2(5^2)$

Practice

49. The factors of 10 are 1, 2, 5 and 10.
51. The factors of 40 are 1, 2, 4, 5, 8, 10, 20 and 40.
53. The factors of 18 are 1, 2, 3, 6, 9 and 18.
55. The factors of 44 are 1, 2, 4, 11, 22 and 44.
57. The factors of 77 are 1, 7, 11 and 77.
59. The factors of 100 are 1, 2, 4, 5, 10, 20, 25, 50 and 100.
61. The prime-factored form of 39 is $3 \cdot 13$.
63. The prime-factored form of 99 is $3^2 \cdot 11$.
65. The prime-factored form of 162 is $2 \cdot 3^4$.
67. The prime-factored form of 220 is $2^2 \cdot 5 \cdot 11$.
69. The prime-factored form of 64 is 2^6.
71. The prime-factored form of 147 is $3 \cdot 7^2$.
73. $3^4 = 3 \cdot 3 \cdot 3 \cdot 3 = 81$
75. $2^5 = 2 \cdot 2 \cdot 2 \cdot 2 \cdot 2 = 32$
77. $12^2 = 144$
79. $8^4 = 8 \cdot 8 \cdot 8 \cdot 8 = 4{,}096$
81. $3^2(2^3) = 3 \cdot 3 \cdot 2 \cdot 2 \cdot 2 = 72$
83. $2^3 \cdot 3^3 \cdot 4^2 = 2 \cdot 2 \cdot 2 \cdot 3 \cdot 3 \cdot 3 \cdot 4 \cdot 4 = 3{,}456$
85. $234^3 = 234 \cdot 234 \cdot 234 = 12{,}812{,}904$
87. $23^2 \cdot 13^3 = 23 \cdot 23 \cdot 13 \cdot 13 \cdot 13 = 1{,}162{,}213$

Applications

89. PERFECT NUMBERS
The factors of 28 are 1, 2, 4, 7, 14, and 28
$1 + 2 + 4 + 7 + 14 = 28$ therefore 28 is a perfect number.

91. LIGHT
From the picture, 1 yd from the bulb the light energy passes through 1 square unit of area, 2 yds from the bulb the light energy passes through $4 = 2^2$ square units of area, 3 yds from the bulb the light energy passes through $9 = 3^2$ square units of area and 4 yds from the bulb the light energy passes through $16 = 4^2$ square units of area.

Writing

93. Answers will vary.
95. Answers will vary

Review

97. 230,999 rounded to the nearest thousand is 231,000.
99. $0 \div 15 = 0$
101. Area of a rectangle is $A = lw$.

Section 1.5 Order of Operations

Vocabulary

1. The grouping symbols ( ) are called **parentheses**, and the symbols [ ] are called **brackets**.
3. To **evaluate** $2+5\cdot4$ means to find its value.

Concepts

5. Three operations need to be performed; square, multiply and subtract.
7. In the numerator, multiplications should be done first. In the denominator, subtraction should be done first.
9. $2\cdot3^2=2\cdot3\cdot3=2\cdot9=18$ and $(2\cdot3)^2=(6)^2=36$

Notation

11. $$\begin{aligned}28-5(2)^2&=28-5(4)\\&=28-20\\&=8\end{aligned}$$

13. $$\begin{aligned}[4(2+7)]-6&=[4(9)]-6\\&=36-6\\&=30\end{aligned}$$

Practice

15. $7+4\cdot5=7+20=27$

17. $2+3(0)=2+0=2$

19. $20-10+5=10+5=15$

21. $25\div5\cdot5=5\cdot5=25$

23. $7(5)-5(6)=35-30=5$

25. $4^2+3^2=16+9=25$

27. $2\cdot3^2=2\cdot9=18$

29. $3+2\cdot3^4\cdot5=3+2\cdot81\cdot5=3+810=813$

31. $5 \cdot 10^3 + 2 \cdot 10^2 + 3 \cdot 10^1 + 9$
$= 5 \cdot 1000 + 2 \cdot 100 + 3 \cdot 10 + 9$
$= 5000 + 200 + 30 + 9$
$= 5{,}239$

33. $3(2)^2 - 4(2) + 12 = 3(4) - 8 + 12 = 12 - 8 + 12 = 4 + 12 = 16$

35. $(8-6)^2 + (4-3)^2 = (2)^2 + (1)^2 = 4 + 1 = 5$

37. $60 - (6 + \frac{40}{8}) = 60 - (6+5) = 60 - (11) = 49$

39. $6 + 2(5+4) == 6 + 2(9) = 6 + 18 = 24$

41. $3 + 5(6-4) = 3 + 5(2) = 3 + 10 = 13$

43. $(7-4)^2 + 1 = (3)^2 + 1 = 9 + 1 = 10$

45. $6^3 - (10+8) = 216 - (18) = 198$

47. $50 - 2(4)^2 = 50 - 2(16) = 50 - 32 = 18$

49. $16^2 - 4(2)(5) = 256 - 40 = 216$

51. $39 - 5(6) + 9 - 1 = 39 - 30 + 9 - 1 = 17$

53. $(18-12)^3 - 5^2 = (6)^3 - 25 = 216 - 25 = 191$

55. $2(10 - 3^2) + 1 = 2(10-9) + 1 = 2(1) + 1 = 2 + 1 = 3$

57. $6 + \frac{25}{5} + 6(3) = 6 + 5 + 18 = 29$

59. $3\left(\frac{18}{3}\right) - 2(2) = 3(6) - 4 = 18 - 4 = 14$

61. $(2 \cdot 6 - 4)^2 = (12-4)^2 = (8)^2 = 64$

63. $4[50 - (3^3 - 5^2)] = 4[50 - (27-25)] = 4[50 - (2)] = 4[48] = 192$

65. $80 - 2[12 - (5+4)] = 80 - 2[12 - (9)] = 80 - 2[3] = 80 - 6 = 74$

67. $2[100 - (5+4)] - 45 = 2[100 - (9)] - 45 = 2[91] - 45 = 182 - 45 = 137$

69. $\frac{10+5}{6-1} = \frac{15}{5} = 3$

71. $\dfrac{5^2+17}{6-2^2}=\dfrac{25+17}{6-4}=\dfrac{42}{2}=21$

73. $\dfrac{(3+5)^2+2}{2(8-5)}=\dfrac{(8)^2+2}{2(3)}=\dfrac{64+2}{6}=\dfrac{66}{6}=11$

75. $\dfrac{(5-3)^2+2}{4^2-(8+2)}=\dfrac{(2)^2+2}{16-(10)}=\dfrac{4+2}{6}=\dfrac{6}{6}=1$

77. $12{,}985-(1{,}800+689)=12{,}985-(2{,}489)=10{,}496$

79. $3{,}245-25(16-12)^2=3{,}245-25(4)^2=3{,}245-25(16)=3{,}245-400=2{,}845$

Applications

81. BUYING GROCERIES

$2(6)+4(2)+2(1)=12+8+2=22$

The total cost of the groceries is $22.

83. BANKING

The total amount of cash being deposited is

$24(1)+0(2)+6(5)+10(10)+12(20)+2(50)+1(100)$

$=24+0+30+100+240+100+100$

$=\$594$

85. SCRABBLE

Brick: $3(3)+1+1+3+3(5)=9+1+1+3+15=29$

Aphid: $3[1+3+4+1+2]=3[11]=33$

87. CLIMATE

The average temperature is $\dfrac{75+80+83+80+77+72+86}{7}=\dfrac{553}{7}=79°$.

89. NATURAL NUMBERS

The average of the first nine natural numbers is

$\dfrac{1+2+3+4+5+6+7+8+9}{9}=\dfrac{45}{9}=5$

91. FAST FOOD

The average number of calories for the group of sandwiches is

$\dfrac{237+289+295+302+303+312+349}{7}=\dfrac{2087}{7}=298$

Writing

93. Answers will vary.
95. Answers will vary.

Review

97. $$\begin{array}{r} 4{,}029 \\ +\,3{,}271 \\ \hline 7{,}300 \end{array}$$

99. $$\begin{array}{r} 417 \\ \times\ \ 23 \\ \hline 1251 \\ 834 \\ \hline 9{,}591 \end{array}$$

Section 1.6 Solving Equations by Addition and Subtraction

Vocabulary

1. An equation is a statement that two expressions are **equal**. An equation contains an = sign.
3. The answer to an equation is called a **solution** or a **root**.
5. **Equivalent** equations have exactly the same solutions.

Concepts

7. If $x = y$ and c is any number, then $x + c = y + c$.
9. In $x + 6 = 10$, the addition of 6 is performed on the variable and subtraction of 6 from both sides would undo the operation and isolate the variable.

Notation

11.
$$x + 8 = 24$$
$$x + 8 - 8 = 24 - 8$$
$$x = 16$$
Check: $x + 8 = 24$
$$16 + 8 \stackrel{?}{=} 24$$
$$24 = 24$$
So 16 is a solution.

Practice

13. $x = 2$ is an equation.
15. $7x < 8$ is not an equation.
17. $x + y = 0$ is an equation.
19. $1 + 1 = 3$ is an equation.
21. 1 is a solution; $1 + 2 = 3$
23. 7 is a solution; $7 - 7 = 0$
25. 5 is not a solution; $8 - 5 \neq 5$
27. 16 is not a solution; $16 + 32 \neq 0$
29. 7 is not a solution; $7 + 7 \neq 7$
31. 0 is a solution; $0 = 0$

33.
$$x - 7 = 3$$
$$x - 7 + 7 = 3 + 7$$
$$x = 10$$

35. $$\begin{aligned} a-2&=5\\ a-2+2&=5+2\\ a&=7 \end{aligned}$$

37. $$\begin{aligned} 1&=b-2\\ 1+2&=b-2+2\\ 3&=b \end{aligned}$$

39. $$\begin{aligned} x-4&=0\\ x-4+4&=0+4\\ x&=4 \end{aligned}$$

41. $$\begin{aligned} y-7&=6\\ y-7+7&=6+7\\ y&=13 \end{aligned}$$

43. $$\begin{aligned} 70&=x-5\\ 70+5&=x-5+5\\ 75&=x \end{aligned}$$

45. $$\begin{aligned} 312&=x-428\\ 312+428&=x-428+428\\ 740&=x \end{aligned}$$

47. $$\begin{aligned} x-117&=222\\ x-117+117&=222+117\\ x&=339 \end{aligned}$$

49. $$\begin{aligned} x+9&=12\\ x+9-9&=12-9\\ x&=3 \end{aligned}$$

51. $$\begin{aligned} y+7&=12\\ y+7-7&=12-7\\ y&=5 \end{aligned}$$

53. $$\begin{aligned} t+19&=28\\ t+19-19&=28-19\\ t&=9 \end{aligned}$$

55. $$\begin{aligned}23+x&=33\\23-23+x&=33-23\\x&=10\end{aligned}$$

57. $$\begin{aligned}5&=4+c\\5-4&=4-4+c\\1&=c\end{aligned}$$

59. $$\begin{aligned}99&=r+43\\99-43&=r+43-43\\56&=r\end{aligned}$$

61. $$\begin{aligned}512&=x+428\\512-428&=x+428-428\\84&=x\end{aligned}$$

63. $$\begin{aligned}x+117&=222\\x+117-117&=222-117\\x&=105\end{aligned}$$

65. $$\begin{aligned}3+x&=7\\3-3+x&=7-3\\x&=4\end{aligned}$$

67. $$\begin{aligned}y-5&=7\\y-5+5&=7+5\\y&=12\end{aligned}$$

69. $$\begin{aligned}4+a&=12\\4-4+a&=12-4\\a&=8\end{aligned}$$

71. $$\begin{aligned}x-13&=34\\x-13+13&=34+13\\x&=47\end{aligned}$$

Applications

73. ARCHAEOLOGY

Analyze the problem

- The manuscript is 1,700 years old.
- The manuscript is 425 years older than the jar.
- We are asked to find the age of the jar.

Form an equation Since we want to find the age of the jar, we can let $x =$ the age of the jar. Now we look for a key word or phrase in the problem.

Key phrase: older than
Translation: add

We can write the age of the manuscript in two ways.

The age of the manuscript is 425 plus the age of the jar.

$1,700 = 425 + x$

Solve the equation

$$1,700 = 425 + x$$
$$1,700 - 425 = 425 + x - 425$$
$$1,275 = x$$

State the conclusion The jar is 1,275 years old.

Check the result If the jar is 1,275 years old, then the manuscript is $1,275 + 425 = 1,700$ years old. The answer checks.

75. ELECTIONS

Let $t =$ the total number of votes received by the three major candidates.

The total number of votes cast is the sum of the number of votes received by the three major candidates.

$t = 47,401,185 + 39,197,469 + 8,085,294$

$t = 94,683,948$ votes

77. PARTY INVITATIONS

Let $s =$ the number of invitations she sent.

The number of invitations sent is the sum of the number delivered and the number lost.

$s = 59 + 3$

$s = 62$ invitations

79. FAST FOOD

Let $b =$ the amount of money the entrepreneur will need to borrow.

The amount of money she will need to borrow is the franchise fee and startup costs minus the amount she has to invest.

$b = 287,000 - 68,500$

$b = \$218,500$

81. CELEBRITY EARNINGS

Let O = the amount that Oprah Winfrey earned in 1998

The amount Oprah earned is the amount the sum of the amount Celine Dion earned and $69 million.

$O = 56,000,000 + 69,000,000$

$O = \$125,000,000$

83. POWER OUTAGE

Let a = the amount the meter must increase to cause the system to shut down.

The amount the meter must increase for the system to shut down is 85 minus the current reading on the meter.

$a = 85 - 60$

$a = 25$ units

85. AUTO REPAIR

Let p = the amount she paid to have her car fixed.

The amount she paid to have her car fixed is the amount that the gas station would have charges minus $29.

$p = 219 - 29$

$p = \$190$

Writing

87. Answers will vary.
89. Answers will vary.
91. Answers will vary.

Review

93. Rounded to the nearest ten 325,784 is 325,780.

95. $2 \cdot 3^2 \cdot 5 = 2 \cdot 9 \cdot 5 = 90$

97. $8 - 2(3) + 1^3 = 8 - 6 + 1 = 3$

Section 1.7 Solving Equations by Division and Multiplication

Vocabulary

1. According to the **division** property of equality, "If equal quantities are divided by the same nonzero quantity, the results will be equal quantities."

Concepts

3. $\frac{6x}{6} = x$

5. If $x = y$, then $\frac{x}{z} = \frac{y}{z}$ $(z \neq 0)$.

7. The variable is being multiplied by 4 and can be undo by dividing by 4.

9. Name the first step in solving
   a. Subtract 5 from both sides
   b. Add 5 to both sides
   c. Divide both sides by 5
   d. Multiply both sides by 5

Notation

11. Complete each solution.

$$3x = 12$$

$$\frac{3x}{3} = \frac{12}{3}$$

$$x = 4$$

Check: $3x = 12$

$$3 \cdot 4 \stackrel{?}{=} 12$$

$$12 = 12$$

So 4 is a solution.

Practice

13. $3x = 3$

$$\frac{3x}{3} = \frac{3}{3}$$

$$x = 1$$

Check: $3x = 3$

$$3 \cdot 1 \stackrel{?}{=} 3$$

$$3 = 3$$

So 1 is a solution.

15. $2x = 192$

$$\frac{2x}{2} = \frac{192}{2}$$

$$x = 96$$

Check: $2x = 192$

$$2 \cdot 96 \stackrel{?}{=} 192$$

$$192 = 192$$

So 96 is a solution.

17. $17y = 51$

$$\frac{17y}{17} = \frac{51}{17}$$

$$y = 3$$

Check: $17y = 51$

$$17 \cdot 3 \stackrel{?}{=} 51$$

$$51 = 51$$

So 3 is a solution.

19. $34y = 204$

$$\frac{34y}{34} = \frac{204}{34}$$

$$y = 6$$

Check: $34y = 204$

$$34 \cdot 6 \stackrel{?}{=} 204$$

$$204 = 204$$

So 6 is a solution.

21. $100 = 100x$

$$\frac{100}{100} = \frac{100x}{100}$$

$$1 = x$$

Check: $100 = 100x$

$$100 \stackrel{?}{=} 100 \cdot 1$$

$$100 = 100$$

So 1 is a solution.

23. $16 = 8r$

$$\frac{16}{16} = \frac{8r}{16}$$

$$2 = r$$

Check: $16 = 8r$

$$16 \stackrel{?}{=} 8 \cdot 2$$

$$16 = 16$$

So 2 is a solution.

25. Solve

$$\frac{x}{7} = 2$$

$$7 \cdot \frac{x}{7} = 2 \cdot 7$$

$$x = 14$$

Check: $\frac{x}{7} = 2$

$$\frac{14}{7} \stackrel{?}{=} 2$$

$$2 = 2$$

So 14 is a solution.

27. Solve

$$\frac{y}{14} = 3$$

$$14 \cdot \frac{y}{14} = 3 \cdot 14$$

$$y = 42$$

Check: $\frac{y}{14} = 3$

$$\frac{42}{14} \stackrel{?}{=} 3$$

$$3 = 3$$

So 42 is a solution.

29. Solve

$$\frac{a}{15} = 5$$

$$15 \cdot \frac{a}{15} = 5 \cdot 15$$

$$a = 75$$

Check: $\frac{a}{15} = 5$

$$\frac{75}{15} \stackrel{?}{=} 5$$

$$5 = 5$$

So 75 is a solution.

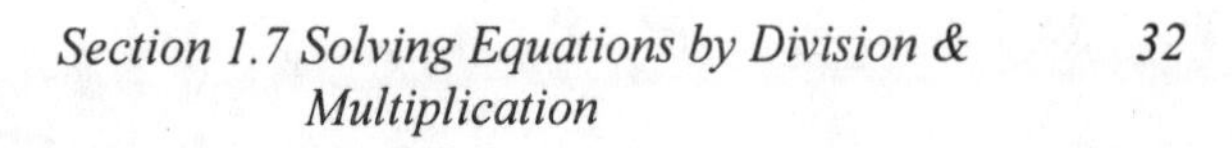

31. Solve

$$\frac{c}{13}=3$$

$$13\cdot\frac{c}{13}=3\cdot 13$$

$$c=39$$

Check: $\frac{c}{13}=3$

$$\frac{39}{13}\overset{?}{=}3$$

$$3=3$$

So 39 is a solution.

33. Solve

$$1=\frac{x}{50}$$

$$50\cdot 1=\frac{x}{50}\cdot 50$$

$$50=x$$

Check: $1=\frac{x}{50}$

$$1\overset{?}{=}\frac{50}{50}$$

$$1=1$$

So 50 is a solution.

35. Solve

$$7=\frac{t}{7}$$

$$7\cdot 7=\frac{x}{7}\cdot 7$$

$$49=x$$

Check: $7=\frac{t}{7}$

$$1\overset{?}{=}\frac{7}{7}$$

$$1=1$$

So 49 is a solution.

37. $9z = 90$

$$\frac{9z}{9} = \frac{90}{9}$$

$$z = 10$$

Check: $9z = 90$

$$9 \cdot 10 \stackrel{?}{=} 90$$

$$90 = 90$$

So 10 is a solution.

39. $7x = 21$

$$\frac{7x}{7} = \frac{21}{7}$$

$$x = 3$$

Check: $7x = 21$

$$7 \cdot 3 \stackrel{?}{=} 21$$

$$21 = 21$$

So 3 is a solution.

41. $86 = 43t$

$$\frac{86}{43} = \frac{43t}{43}$$

$$t = 2$$

Check: $86 = 43t$

$$86 \stackrel{?}{=} 43 \cdot 2$$

$$86 = 86$$

So 2 is a solution.

43. $21s = 21$

$$\frac{21s}{21} = \frac{21}{21}$$

$$s = 1$$

Check: $21s = 21$

$$21 \cdot 1 \stackrel{?}{=} 21$$

$$21 = 21$$

So 1 is a solution.

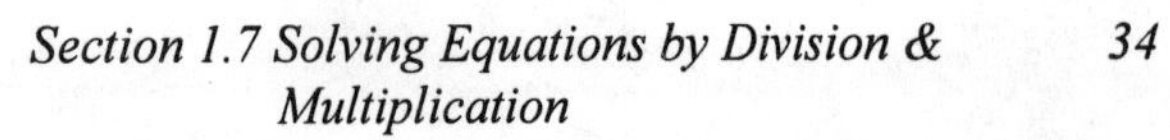

45. Solve

$$\frac{d}{20}=2$$

$$20\cdot\frac{d}{20}=2\cdot 20$$

$$d=40$$

Check: $\frac{d}{20}=2$

$$\frac{40}{20}\stackrel{?}{=}2$$

$$2=2$$

So 40 is a solution.

47. Solve

$$400=\frac{t}{3}$$

$$3\cdot 400=\frac{t}{3}\cdot 3$$

$$1200=t$$

Check: $400=\frac{t}{3}$

$$400\stackrel{?}{=}\frac{1200}{3}$$

$$400=400$$

So 39 is a solution.

Applications

49. NOBEL PRIZE

Analyze the problem

- 3 people shared the cash award.
- Each person received $318,500.
- We are asked to find the Nobel Prize cash award.

Form an equation Since we want to find what the Nobel Prize cash award was, we let $c =$ the Nobel Prize cash award. To form an equation, we look for a key word or phrase in the problem.

Key phrase: shared the prize money

Translation: divide

We can now form the equation.

The Nobel Prize cash award divided by the number of recipients was $318,500.

$x \div 3 = \$318{,}500$

Solve the equation

$$\frac{x}{3} = 318{,}500$$

$$3 \cdot \frac{x}{3} = 3 \cdot 318{,}500$$

$$x = 955{,}500$$

State the conclusion The Nobel Prize cash award was $955,500.

Check the result If we divide the Nobel Prize award by 3, we have

$\frac{\$955{,}500}{3} = \$318{,}500$. This was the amount each person received. The answer checks.

51. SPEED READING

Let $s =$ the speed Alicia can expect to read after taking the classes.

$s = 3 \cdot 130$

$s = 390$ words a minute

53. STAMPS

Let $r =$ the number of rows on each sheet.

$r = \frac{112}{8}$

$r = 14$ rows of stamps per sheet

55. PHYSICAL EDUCATION

Let $s =$ the number of students in the PE class.

$s = 3 \cdot 32$

$s = 96$ students

57. ANIMAL SHELTER

Let $c =$ the number of calls received each day.

$c = 4 \cdot 8$

$c = 32$ calls per day

59. GRAVITY

Let $w =$ the weight the scale would register.

$w = \frac{330}{6}$

$w = 55$ pounds

Writing

61. Answers will vary.
63. Answers will vary.

Review

65. $P = 2l + 2w$

$P = 2(8) + 2(16)$

$P = 48$ cm

67. The prime factorization of 120 is $2^3 \cdot 3 \cdot 5$.

69. $3^2 \cdot 2^3 = 9 \cdot 8 = 72$

71. FUEL ECONOMY

The average city mileage is $\frac{24 + 22 + 28 + 29 + 27}{5} = 26\text{ mph}$.

Chapter 1 Key Concept

1. Let $x =$ the monthly cost to lease the van.

3. Let $x =$ the width of the field.

5. Let $x =$ the distance traveled by the motorist.

7. $a+b=b+a$

9. $\frac{b}{1}=b$

11. $n-1<n$

13. $(r+s)+t=r+(s+t)$

Chapter 1 Review

Section 1.1

1. Graph each set
   a. Natural numbers less than 5

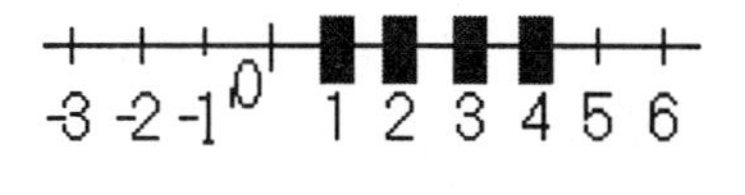

   b. Whole numbers between 0 and 3

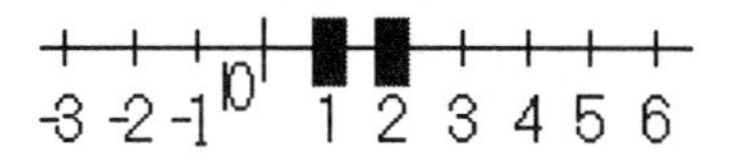

3. Consider 2,365,720
   a. 6 is in the ten thousands column
   b. 7 is in the hundreds column

5. Write in standard notation.
   a. 3,207
   b. 23,253,412
   c. 16,000,000,000

7. Round 2,507,348
   a. 2,507,300
   b. 2,510,000
   c. 2,507,350
   d. 2,500,000

Section 1.2

9. Do each addition.
   a. $\begin{array}{r} 236 \\ +282 \\ \hline 518 \end{array}$
   b. $\begin{array}{r} 5,345 \\ +\ 655 \\ \hline 6,000 \end{array}$
   c. $135 + 213 + 615 + 47 = 135 + 213 + 662 = 135 + 875 = 1,010$
   d. $4,447 + 7,478 + 13,061 = 4,447 + 20,539 = 24,986$

11. $P = 4s$

$P = 4(24) = 96\text{in}$

13. Translate each figure.
   a. $4 + 2 = 6$
   b. $5 - 2 = 3$

15. SAVINGS ACCOUNTS

$931 + 271 - 37 - 380 = 931 + 271 - 417 = 1202 - 417 = 785$

The final balance is $785.

Section 1.3

17. Do each multiplication.
   a. $8 \cdot 7 = 56$
   b. $7(8) = 56$
   c. $8 \cdot 0 = 0$
   d. $7 \cdot 1 = 7$
   e. $10 \cdot 8 \cdot 7 = 10 \cdot 56 = 560$
   f. $5 \cdot (7 \cdot 6) = 5 \cdot (42) = 210$

19. State the property.
   a. Associative property of multiplication
   b. Commutative property of multiplication

21. HORSESHOES

$Perimeter = 2l + 2w = 2(48) + 2(6) = 96 + 12 = 108 \text{ ft}$

$Area = lw = 48(6) = 288 \text{ ft}^2$

23. Do each division.
   a. $\frac{6}{3} = 2$
   b. $\frac{15}{1} = 15$
   c. $73 \div 0$ is undefined
   d. $\frac{0}{8} = 0$
   e. $357 \div 17 = 21$
   f. $1,443 \div 39 = 37$

g. $\begin{array}{r} 19 \\ 21\overline{)405} \\ \underline{21} \\ 195 \\ \underline{189} \\ 6 \end{array}$ Answer: 19 R6

h. $\begin{array}{r} 23 \\ 54\overline{)1269} \\ \underline{108} \\ 189 \\ \underline{162} \\ 27 \end{array}$ Answer: 23 R27

25. COPIES

$\frac{84}{3} = 28$

There were 28 copies of the test made.

Section 1.4

27. Identify
   a. 31 is prime
   b. 100 is composite
   c. 1 is neither
   d. 0 is neither
   e. 125 is composite
   f. 47 is prime

29. Find the prime factorization
   a. The prime factorization of 42 is $2 \cdot 3 \cdot 7$.
   b. The prime factorization of 375 is $3 \cdot 5^3$.

31. Evaluate
   a. $5^3 = 125$
   b. $11^2 = 121$
   c. $2^3 \cdot 5^2 = 8 \cdot 25 = 200$
   d. $2^2 \cdot 3^3 \cdot 5^2 = 4 \cdot 27 \cdot 25 = 2,700$

Section 1.5

33. DICE GAME

$$3(6)+2(5)=18+10=28$$

Section 1.6

35. Tell whether the given number is a solution.

a. 5 is not a solution.

$$x+2=13$$
$$5+2\stackrel{?}{=}13$$
$$7\neq 13$$

b. 4 is a solution.

$$x-3=1$$
$$4-3\stackrel{?}{=}1$$
$$1=1$$

37. Solve

a.

$$x-7=2$$
$$x-7+7=2+7$$
$$x=9$$

Check:

$$9-7=2$$

b.

$$x-11=20$$
$$x-11+11=20+11$$
$$x=31$$

Check:

$$31-11=20$$

c.

$$225=y-115$$
$$225+115=y-115+115$$
$$340=y$$

Check:

$$225=340-115$$

d.

$$101 = p - 32$$
$$101 + 32 = p - 32 + 32$$
$$133 = p$$

Check:

$$101 = 133 - 32$$

e.

$$x + 9 = 18$$
$$x + 9 - 9 = 18 - 9$$
$$x = 9$$

Check:

$$9 + 9 = 18$$

f.

$$b + 12 = 26$$
$$b + 12 - 12 = 26 - 12$$
$$b = 14$$

Check:

$$14 + 12 = 26$$

g.

$$175 = p + 55$$
$$175 - 55 = p + 55 - 55$$
$$120 = p$$

Check:

$$175 = 120 + 55$$

h.

$$212 = m + 207$$
$$212 - 207 = m + 207 - 207$$
$$5 = m$$

Check:

$$212 = 5 + 207$$

i.

$$x - 7 = 0$$
$$x - 7 + 7 = 0 + 7$$
$$x = 7$$

Check:

$$7 - 7 = 0$$

j. $x + 15 = 1{,}000$

$$x + 15 - 15 = 1{,}000 - 15$$

$$x = 985$$

Check:

$$985 + 15 = 1{,}000$$

39. DOCTOR'S PATIENTS

Let $p =$ the number of patients originally.

$$p - 13 = 172$$

$$p - 13 + 13 = 172 + 13$$

$$p = 185$$

The doctor originally had 185 patients.

Section 1.7

41. CARPENTRY

Let $l =$ the length of each piece.

$$3l = 72$$

$$\frac{3l}{3} = \frac{72}{3}$$

$$l = 24$$

Each piece will be 24 inches long.

Chapter 1 Test

1. Graph the whole numbers less than 5.

-3 -2 -1 1 2 3 4 5 6 7

3. 7,507

5. Bar Graph

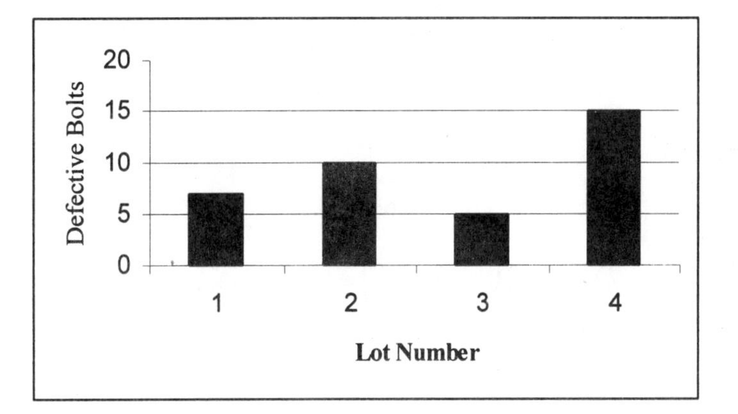

7. $15 > 10$

9. $327 + 435 + 123 + 606 = 327 + 435 + 729 = 327 + 1{,}164 = 1{,}491$

11.
$$\begin{array}{r} 44{,}526 \\ +\,13{,}579 \\ \hline 58{,}105 \end{array}$$

13. STOCKS
Let $p =$ the price on Thursday.
$p = 73 + 12 - 9 = 73 + 3 = \$76$

15.
$$\begin{array}{r} 53 \\ \times\ 8 \\ \hline 424 \end{array}$$

17.
$$\begin{array}{r} 72 \\ 63\overline{)4{,}536} \\ \underline{441} \\ 126 \\ \underline{126} \\ 0 \end{array}$$

19. FURNITURE SALE

$$\begin{aligned} Perimeter &= 2l + 2w \\ &= 2(105) + 2(75) \\ &= 210 + 150 \\ &= 360\text{ ft} \end{aligned}$$

The advertisement asks for square feet, which means we need to find the area.

$$\begin{aligned} Area &= lw \\ &= 105(75) \\ &= 7{,}875 \text{ square feet} \end{aligned}$$

21. COLLECTIBLES

There are $12 \cdot 24 \cdot 12 = 12 \cdot 288 = 3{,}456$ cards in a case.

23. $3 \cdot 4^2 - 2^2 = 3 \cdot 16 - 4 = 48 - 4 = 44$

25. $10 + 2[12 - 2(6-4)] = 10 + 2[12 - 2(2)] = 10 + 2[12 - 4] = 10 + 2[8] = 10 + 16 = 26$

27. $x + 13 = 16$

$$\begin{aligned} 3 + 13 &\stackrel{?}{=} 16 \\ 16 &= 16 \end{aligned}$$

3 is a solution to the equation.

29.
$$\begin{aligned} y - 12 &= 18 \\ y - 12 + 12 &= 18 + 12 \\ y &= 30 \end{aligned}$$

Check:

$$30 - 12 = 18$$

31.
$$\begin{aligned} \frac{q}{3} &= 27 \\ 3 \cdot \frac{q}{3} &= 27 \cdot 3 \\ q &= 81 \end{aligned}$$

Check:

$$\frac{81}{3} = 27$$

33. LIBRARY

Let b = the age of the building.

$$b + 6 = 200$$

$$b + 6 - 6 = 200 - 6$$

$$b = 194 \text{ years old}$$

Section 2.1 An Introduction to the Integers

Vocabulary

1. **Negative** numbers are less than 0.
3. Numbers can be represented by points equally spaced on a **number** line.
5. The symbols > and < are called **inequality** symbols.
7. Two numbers on a number line that are the same distance from zero, but on opposite sides of the origin, are called **opposites**.

Concepts

9. The values of numbers get smaller as we move to the left on a number line.

11. Every integer has an opposite.

13. Fifteen subtract eight is represented with a subtraction sign.

15. $15 > 12$

17. a. -225
    b. -10
    c. -3
    d. $-12,000$

19. Negative four is three units to the right of negative seven.
21. Negative eight and two are five units from three on a number line.
23. Negative seven is closer to negative three; 2 is 5 units away and -7 is 4 units away.
25. Three examples could include using $-$ as a subtraction sign, as a negative or to show the opposite. For example: $3-1, -9, -(-3)$.

Notation

27. a. $-(-8)$
    b. $|-8|$
    c. $8-8$
    d. $-|-8|$

Practice

29. $|9| = 9$

31. $|-8| = 8$

33. $|-14| = 14$

35. $-|20| = -20$

37. $-|-6| = -6$

39. $|203| = 203$

41 $-0 = 0$

43. $-(-11) = 11$

45. $-(-4) = 4$

47. $-(-1,201) = 1,201$

49. $\{-3, -1, 0, 3, 4\}$

-3 -2 -1 0 1 2 3 4

51. $\{-(-3), -(5), |-2|\}$

-5 -4 -3 -2 -1 0 1 2 3

53. $-5 < 5$

55. $-12 < -6$

57. $-10 > -11$

59. $|-2| > 0$

61. $-1,255 < -(-1,254)$

63. $-|-3| < 4$

Applications

65. FLIGHT OF A BALL

Time (sec)	Position
1	2
2	3
3	2
4	0
5	-3
6	-7

67. TECHNOLOGY

The peaks occur at 2, 4, and 0. The valleys occur at –3, -5 and –2.

69. GOLF

a. –1, or one below par was most often shot on this hole.

b. –3, 3 below par was the best score on this hole.

c. This appears too easy as most scores are below par.

71. WEATHER MAP

a. The temperature range is negative ten to negative twenty.
b. Chicago will be approximately ten degrees colder than Denver.
c. The coldest it should get in Seattle is ten degrees.

73. HISTORICAL TIME LINE

a. The basic unit was 200 years.
b. A.D. could be thought of as positive numbers
c. B.C. could be thought of as negative numbers.
d. The birth of Christ distinguishes positive from negative.

75. LINE GRAPH

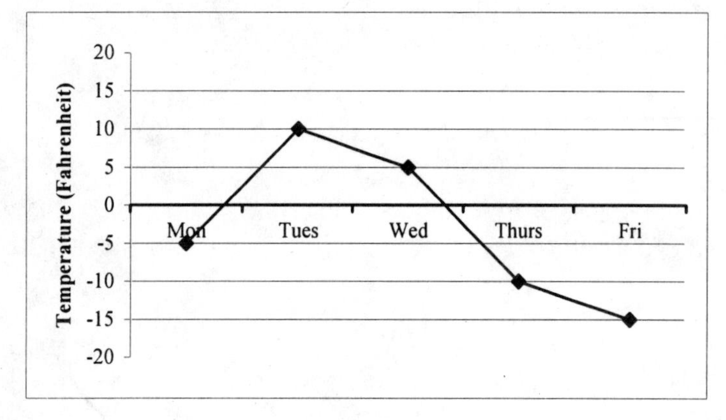

Writing

77. Answers may vary
79. Answers may vary
81. Answers may vary

Review

83. 23,500

85. $2x = 34$

$x = 17$

87. This illustrates the associative property of multiplication.

Section 2.2 Addition of Integers

Vocabulary

1. When 0 is added to a number, the number remains the same. We call 0 the additive **identity**.

Concepts

3. $-3+6=3$
5. $-5+3=-2$

7. a. The sum of two positive integers is always positive.
   b. The sum of two negative integers is always negative.

9. a. $|-7|=7$
   b. $|10|=10$

11. To add two integers with unlike signs, **subtract** their absolute values, the smaller from the larger. Then attach to that result the sign of the number with the **larger** absolute value.

Notation

13. $-16+(-2)+(-1)=-18+(-1)$
    $=-19$

15. $(-3+8)+(-3)=5+(-3)$
    $=2$

17. $-6+-5$ could be written as $-6+(-5)$

Practice

19. -11 additive inverse 11
21. -23 additive inverse 23
23. 0 additive inverse 0
25. 99 additive inverse -99
27. $-6+(-3)=-9$
29. $-5+(-5)=-10$
31. $-6+7=1$
33. $-15+8=-7$
35. $20+(-40)=-20$
37. $30+(-15)=15$
39. $-1+9=8$

41. $-7+9=2$
43. $5+(-15)=-10$
45. $24+(-15)=9$
47. $35+(-27)=8$
49. $24+(-45)=-21$
51. $-2+6+(-1)=4+(-1)=3$
53. $-9+1+(-2)=-8+(-2)=-10$
55. $6+(-4)+(-13)+7=2+(-13)+7=-11+7=-4$
57. $9+(-3)+5+(-4)=6+5+(-4)=11+(-4)=7$
59. $-6+(-7)+(-8)=-13+(-8)=-21$
61. $-7+0=-7$
63. $9+0=9$
65. $-4+4=0$
67. $2+(-2)=0$
69. $5+(-5)=0$
71. $2+(-10+8)=2+(-2)=0$
73. $(-4+8)+(-11+4)=4+(-7)=-3$
75. $[-3+(-4)]+(-5+2)=-7+(-3)=-10$
77. $[6+(-4)]+[8+(-11)]=2+(-3)=-1$
79. $-2+[-8+(-7)]=-2+(-15)=-17$
81. $789+(-9{,}135)=-8{,}346$
83. $-675+(-456)+99=-1{,}131+99=-1{,}032$

Applications

85. G FORCES
$1G+2G=3G$
$1G+(-4G)=-3G$

87. CASH FLOW
$900-(450+380)=\$70$ shortfall each month.

89. MEDICAL QUESTIONNAIRE
This patient's risk is $-1+(-3)+(-2)+3+2+2=-6+7=1$ which places him in the 2% risk category.

91. ATOMS

The net charge of atom (a) is $(-1)+(-1)+(-1)+(-1)+1+1+1=-1$

The net charge of atom (b) is $(-1)+(-1)+(-1)+(-1)+1+1+1+1=0$

93. FLOODING

Consider four feet under flood stage as negative four so the river is now $-4+11=7$ feet above flood stage.

95. FILM PROFITS

The company earned a profit of

10 million $+(-5$ million$)+15$ million $+(-10$ million$)=\$10$ million

Writing

97. Answers may vary
99. Answers may vary

Review

101. The area of this rectangle is $5(3)=15\text{ ft}^2$.

103. $x-7=20$

$x=27$

105. The prime factorization of 125 is $125=25 \cdot 5=5^2 \cdot 5=5^3$

Section 2.3 Subtraction of Integers

Vocabulary

1. The answer to a subtraction problem is called the **difference**.

Concepts

3. **Subtraction** is the same as adding the opposite of the number to be subtracted.
5. Subtracting –6 is the same as adding 6.
7. For any numbers x and y, $x-y=x+(-y)$.
9. After using parentheses as grouping symbols, if another set of grouping symbols is needed, we use **brackets**.
11. Negative eight minus negative four is $-8-(-4)$
13. The distance can be found using $3-(-4)=3+4=7$ units
15. $8-3=5$ and $3-8=3+(-8)=-5$ therefore they are not the same.

Notation

17. $1-3-(-2)=1+(-3)+2$
$=-2+2$
$=0$

19. $(-8-2)-(-6)=[-8+(-2)]-(-6)$
$=-10-(-6)$
$=-10+6$
$=-4$

Practice

21. $8-(-1)=8+1=9$
23. $-4-9=-4+(-9)=-13$
25. $-5-5=-5+(-5)=-10$
27. $-5-(-4)=-5+4=-1$
29. $-1-(-1)=-1+1=0$
31. $-2-(-10)=-2+10=8$
33. $0-(-5)=0+5=5$
35. $0-4=0+(-4)=-4$

37. $-2-2=-2+(-2)=-4$

39. $-10-10=-10+(-10)=-20$

41. $9-9=9+(-9)=0$

43. $-3-(-3)=-3+3=0$

45. $-4-(-4)-15=-4+4-15$
$=0-15$
$=-15$

47. $-3-3-3=-3+(-3)+(-3)=-9$

49. $5-9-(-7)=5+(-9)+7$
$=-4+7$
$=3$

51. $10-9-(-8)=1+8$
$=9$

53. $-1-(-3)-4=-1+3+(-4)$
$=2+(-4)$
$=-2$

55. $-5-8-(-3)=-5+(-8)+3$
$=-13+3$
$=-10$

57. $(-6-5)-3=[-6+(-5)]-3$
$=-11+(-3)$
$=-14$

59. $(6-4)-(1-2)=2-[1+(-2)]$
$=2-(-1)$
$=2+1$
$=3$

61. $-9-(6-7)=-9-[6+(-7)]$
$=-9-(-1)$
$=-9+1$
$=-8$

63. $-8-[4-(-6)]=-8-[4+6]$
$$\begin{aligned}&=-8-10\\&=-8+(-10)\\&=-18\end{aligned}$$

65. $[-4+(-8)]-(-6)=-12+6=-6$

67. $7-(-3)=7+3=10$

69. $-10-(-6)=-10+6=-4$

71. $-1,557-890=-1,557+(-890)=-2,447$

73. $20,007-(-496)=20,007+496=20,503$

75. $-162-(-789)-2,303=-162+789-2,303$
$$\begin{aligned}&=627+(-2,303)\\&=-1,676\end{aligned}$$

Applications

77. SCUBA DIVING
$-50-70=-50+(-70)=-120$ feet

79. READING PROGRAM
The reading scores improved by
$-23+x=-7$
$x=16$ points

81. AMPERAGE
$5-7-6=5+(-7)+(-6)=-2+(-6)=-8$

83. GEOGRAPHY
$-283-(-1,290)=-283+1,290=1,007$ feet

85. FOOTBALL
After the third play $-1+(-6)+(-5)+8=-12+8=-4$ yards

87. DIVING

a.

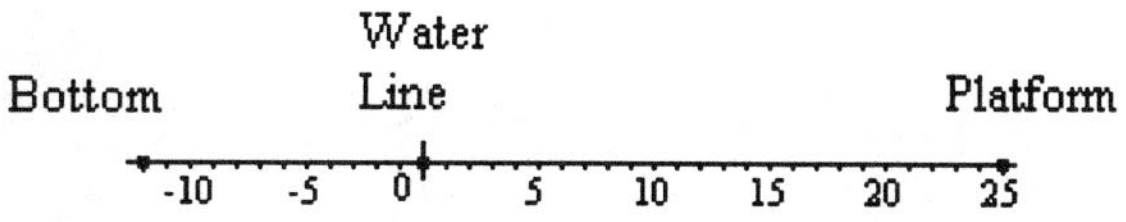

b. The total dive length $25-(-12)=25+12=37$ feet

89. CHECKING ACCOUNT

$$\begin{aligned}1{,}303-676-121-750&=627-121-750\\&=506-750\\&=506+(-750)\\&=-244\end{aligned}$$

Michael will be overdrawn by \$244.

Writing

91. Answers may vary
93. Answers may vary

Review

95. $5x=15 \rightarrow x=3$
97. Factors of twenty are $\{1,2,4,5,10,20\}$
99. To make twelve cans of orange juice you need $12(13)=156$ oranges
101. $4{,}502=4(1{,}000)+5(100)+0(10)+2(1)$

Section 2.4 Multiplication of Integers

Vocabulary

1. In the multiplication $-5(-4)$, the integers –5 and –4 which are being multiplied, are called **factors**. The answer, 20, is called the **product**.
3. In the expression -3^5, **3** is the base and 5 is the **exponent**.

Concepts

5. The product of two integers with **unlike** signs is negative.
7. The **commutative** property of multiplication implies that $-2(-3) = -3(-2)$.
9. $-1(9) = -9$; In general the result of multiplication by negative one is the opposite of the number being multiplied.
11. The four possible combinations, not including multiplication by zero, are: positive(positive); positive(negative); negative(positive); negative(negative).

13. a. negative result
    b. positive result

15. a. $|-3| = 3$
    b. $|12| = 12$
    c. $|-5| = 5$
    d. $|9| = 9$
    e. $|10| = 10$
    f. $|-25| = 25$

17.

Problem	*Negative Factors*	*Answer*
$-2(-2)$	2	4
$-2(-2)(-2)(-2)$	4	16
$-2(-2)(-2)(-2)(-2)(-2)$	6	64

The product of an **even** number of negative integers is positive.

Notation

19. $-3(-2)(-4) = 6(-4)$
    $= -24$

21. Accepted practice would have parenthesis about the negative five; $-6(-5)$.

Practice

23. $-9(-6)=54$
25. $-3 \cdot 5=-15$
27. $12(-3)=-36$
29. $(-8)(-7)=56$
31. $(-2)10=-20$
33. $-40 \cdot 3=-120$
35. $-8(0)=0$
37. $-1(-6)=6$
39. $-7(-1)=7$
41. $1(-23)=-23$
43. $-6(-4)(-2)=24(-2)=-48$
45. $5(-2)(-4)=-10(-4)=40$
47. $2(3)(-5)=6(-5)=-30$
49. $6(-5)(2)=-30(2)=-60$
51. $(-1)(-1)(-1)=1(-1)=-1$
53. $-2(-3)(3)(-1)=6(3)(-1)=18(-1)=-18$
55. $3(-4)(0)=0$
57. $-2(0)(-10)=0$
59. $-6(-10)=60$
61. $(-4)^2=16$
63. $(-5)^3=-125$
65. $(-2)^3=-8$
67. $(-9)^2=81$
69. $(-1)^5=-1$
71. $(-1)^8=1$
73. $(-7)^2=49$ and $-7^2=-49$
75. $(-12)^2=144$ and $-12^2=-144$
77. $-76(787)=-59,812$
79. $(-81)^4=43,046,721$
81. $(-32)(-12)(-67)=384(-67)=-25,728$
83. $(-25)^4=390,625$

Applications

85. DIETING
    a. Plan #1 has an expected weight loss of $10(3) = 30$ pounds, plan #2 has an expected weight loss of $14(2) = 28$ pounds.
    b. Plan #1 is the plan for the most weight loss but may not be chosen since the exercise time is double that of plan #2.

87. MAGNIFICATION
    a. The high value is two and the low value is negative three.
    b. The new high would be four and the new low would be negative six.

89. TEMPERATURE CHANGES
    The total temperature change would be $-4(5) = -20°$

91. EROSION
    During the next decade the levy will decrease $-2(10) = -20$ feet

93. WOMEN'S NATIONAL BASKETBALL ASSOCIATION
    The financial loss would be $11{,}906(-3) = -\$35{,}718$

Writing

95. Answers may vary
97. Answers may vary

Review

99. $3^2 \cdot 5 = 9(5) = 45$
101. The enrollment increase was $12{,}300 - 10{,}200 = 2{,}100$ students
103. < means less than

Section 2.5 Division of Integers

Vocabulary

1. In $\frac{-27}{3}=-9$, the number -9 is called the **quotient**, and the number 3 is the **divisor**.
3. The **absolute value** of a number is the distance between it and 0 on the number line.
5. The quotient of two negative integers is **positive**.

Concepts

7. $5(-5)=-25$
9. $0(?)=-6$
11. There are two related division statements, $\frac{-20}{5}=-4$ or $\frac{-20}{-4}=5$
13. a. This is always true.
    b. This is sometimes true.
    c. This is always true.

Practice

15. $\frac{-14}{2}=-7$
17. $\frac{-8}{-4}=2$
19. $\frac{-25}{-5}=5$
21. $\frac{-45}{-15}=3$
23. $\frac{40}{-2}=-20$
25. $\frac{50}{-25}=-2$
27. $\frac{0}{-16}=0$
29. $\frac{-6}{0}=$ undefined

31. $\frac{-5}{1} = -5$

33. $-5 \div (-5) = 1$

35. $\frac{-9}{9} = -1$

37. $\frac{-10}{-1} = 10$

39. $\frac{-100}{25} = -4$

41. $\frac{75}{-25} = -3$

43. $\frac{-500}{-100} = 5$

45. $\frac{-200}{50} = -4$

47. $\frac{-45}{9} = -5$

49. $\frac{8}{-2} = -4$

51. $\frac{-13,550}{25} = -542$

53. $\frac{272}{-17} = -16$

Applications

55. TEMPERATURE DROP

The average change in this time span was $\frac{-20}{5} = -4°$.

57. SUBMARINE DIVE

$\frac{-3,000}{3} = -1,000$ feet describes how deep each of the three dives will be.

59. BASEBALL TRADE

To make up being twelve games behind or -12 the team hopes to finish the season six games back or $\frac{-12}{2}=-6$.

.61. PRICE MARKDOWN

She can mark down each pair of jeans by $\frac{-300}{20}=-\$15$.

63. PAY CUT

$\frac{-9{,}135{,}000}{5{,}250}=-1{,}740$, each employee will experience a $-\$1{,}740$ pay cut.

Writing

65. Answers may vary
67. Answers may vary

Review

69. $3\left(\frac{18}{3}\right)^2-2(2)=3(36)-4=104$

71. The prime factorization of $210=21\cdot10=3\cdot7\cdot2\cdot5=2\cdot3\cdot5\cdot7$

73. $99=r+43$

$56=r$

75. $3^4=81$

Section 2.6 Order of Operations and Estimation

Vocabulary

1. When asked to evaluate expressions containing more than one operation, we should apply the rules for the **order** of operations.
3. Absolute value symbols, parentheses, and brackets are types of **grouping** symbols.

Concepts

5. Three operations need to be performed to evaluate this expression. First, the power, second multiplication and finally the subtraction.
7. In the numerator the multiplication should be performed first. In the denominator the subtraction should be performed first.
9. The difference of these two expressions is the base -3^2 has base 3 whereas $(-3)^2$ has base (-3).

Notation

11. $$\begin{aligned} -8-5(-2)^2 &= -8-5(4) \\ &= -8-20 \\ &= -8+(-20) \\ &= -28 \end{aligned}$$

13. $$\begin{aligned} [-4(2+7)]-6 &= [-4(9)]-6 \\ &= -36-6 \\ &= -42 \end{aligned}$$

Practice

15. $(-3)^2-4^2=9-16=-7$

17. $$\begin{aligned} 3^2-4(-2)(-1) &= 9+(-4)(-2)(-1) \\ &= 9+8(-1) \\ &= 1 \end{aligned}$$

19. $(2-5)(5+2)=-3(7)=-21$

21. $-10-2^2=-10-4=-14$

23. $\dfrac{-6-8}{2}=\dfrac{-14}{2}=-7$

25. $\frac{-5-5}{2}=\frac{-10}{2}=-5$

27. $-12\div(-2)2=6\bullet 2=12$

29. $-16-4\div(-2)=-16-(-2)=-16+2=-14$

31. $|-5(-6)|=|30|=30$

33. $|-4-(-6)|=|2|=2$

35. $5|3|=5(3)=15$

37. $-6|-7|=-6(7)=-42$

39. $(7-5)^2-(1-4)^2=2^2-(-3)^2=4-9=-5$

41. $-1(2^2-2+1^2)=-1(4-2+1)=-1(3)=-3$

43. $-50-2(-3)^3=-50-2(-27)=-50+54=4$

45. $-6^2+6^2=-36+36=0$

47. $3\left(\frac{-18}{3}\right)-2(-2)=3(-6)+4=-18+4=-14$

49. $6+\frac{25}{-5}+6\bullet 3=6+(-5)+18=19$

51. $\frac{1-3^2}{-2}=\frac{1-9}{-2}=\frac{-8}{-2}=4$

53. $\frac{-4(-5)-2}{-6}=\frac{20-2}{-6}=\frac{18}{-6}=-3$

55. $-3\left(\frac{32}{-4}\right)-1(-1)^5=-3(-8)+1=24+1=25$

57. $6(2^3)(-1)=6(8)(-1)=48(-1)=-48$

59. $2+3[5-(1-10)]=2+3[5+9]=2+3(14)=44$

61. $-7(2-3\bullet 5)=-7(2-15)=-7(-13)=91$

63. $-[6-(1-4)^2]=-[6-(-3)^2]$
$=-[6-9]$
$=-(-3)$
$=3$

65. $15+(-3\cdot4-8)=15+[-12+(-8)]$
$=15+(-20)$
$=-5$

67. $|-3\cdot4+(-5)|=|-12+(-5)|=|-17|=17$

69. $|(-5)^2-2\cdot7|=|25-14|=|11|=11$

71. $-2+|6-4^2|=-2+|6-16|$
$=-2+|-10|$
$=-2+10$
$=8$

73. $2|1-8|\cdot|-8|=2|-7|\cdot8=2(7)(8)=14(8)=112$

75. $-2(-34)^2-(-605)=-2(1{,}156)+605=-2{,}312+605=-1{,}707$

77. $-60-\frac{1{,}620}{-36}=-60-(-45)=-60+45=-15$

79. $-379+(-103)+287\approx-200$

81. $-39\cdot8\approx-320$

83. $-3{,}887+(-5{,}106)\approx-9{,}000$

85. $\frac{6{,}267}{-5}\approx-1{,}200$

Applications

87. TESTING
Twelve correct, three wrong and five unanswered results in a score of
$12(3)+3(-4)+5(-1)=36+(-12)+(-5)=19$

89. SCOUTING REPORT
The average gain was $\frac{16+10+(-2)+0+4+(-4)+66+(-2)}{8}=\frac{88}{8}=11$ yards

91. OIL PRICES
There appears to be about a 102-cent loss for the week.
$(91-68)+(-47-91)+[-22-(-47)]$
$=23+(-138)+25+(-12)$
$=-102$

Writing

93. Answers may vary
95. Answers may vary

Review

97. $8 = 2x$

$$4 = x$$

99. The perimeter of a rectangle is found be summing the lengths of all sides.
101. The elevator is not overloaded; the weight of the passengers is $7(140) = 980$ pounds.

Section 2.7 Solving Equations Involving Integers

Vocabulary

1. To **solve** an equation, we isolate the variable on one side of the = sign.

Concepts

3. $\frac{x(-3)}{-3} = x$

5. a. $x + 3 - 3 = 10 - 3$
   b. $x + 3 + (-3) = 10 + (-3)$

7. a. $-2x = -100$, multiplication by -2
   b. $-6 + x = -9$ addition of -6
   c. $-4x - 8 = 12$ multiplication by -4 and subtraction of 8
   d. $-1 = -6 + (-5x)$ multiplication by -5 and addition of -6

9. When solving the equation $-4 + t = -8 - 2$, it is best to **simplify** the right-hand side of the equation first before undoing any operations performed on the variable.

11. When solving an equation, we isolate the variable by undoing the operations performed on it in the **opposite** order.

13. a. $-2x - 3 = -19$, undo the subtraction of 3 first.
    b. $-6 + \frac{h}{-3} = -14$, undo the addition of -6 first.

Notation

15.
$$\begin{aligned} y + (-7) &= -16 + 3 \\ y + (-7) &= -13 \\ y + (-7) + 7 &= -13 + 7 \\ y &= -6 \end{aligned}$$

17.
$$\begin{aligned} -13 &= -4y - 1 \\ -13 + 1 &= -4y - 1 + 1 \\ -12 &= -4y \\ \frac{-12}{-4} &= \frac{-4y}{-4} \\ 3 &= y \\ y &= 3 \end{aligned}$$

19. $-10x$ means $-10 \cdot x$

Practice

21. $-3x-4=2;-2$

$$-3(-2)-4\overset{?}{=}2$$
$$2=2$$

-2 is a solution

23. $-x+8=-4;4$

$$-4+8\overset{?}{=}-4$$
$$4\neq-4$$

4 is not a solution

25.
$$x+6=-12$$
$$x=-12-6$$
$$x=-18$$
check
$$-18+6=-12$$

27.
$$-6+m=-20$$
$$m=-20+6$$
$$m=-14$$
check
$$-6+(-14)=-20$$

29.
$$-5+3=-7+f$$
$$-2=-7+f$$
$$-2+7=f$$
$$5=f$$
check
$$-2=-7+5$$

31.
$$h-8=-9$$
$$h=-9+8$$
$$h=-1$$
check
$$-1-8=-9$$

33. $0 = y + 9$

$-9 = y$

check

$0 = -9 + 9$

35. $r - (-7) = -1 - 6$

$r + 7 = -7$

$r = -7 - 7$

$r = -14$

check

$-14 - (-7) = -7$

37. $t - 4 = -8 - (-2)$

$t - 4 = -6$

$t = -6 + 4$

$t = -2$

check

$-2 - 4 = -6$

39. $x - 5 = -5$

$x = -5 + 5$

$x = 0$

check

$0 - 5 = -5$

41. $-2s = 16$

$s = \frac{16}{-2}$

$s = -8$

check

$-2(-8) = 16$

43. $-5t = -25$

$s = \frac{-25}{-5}$

$s = 5$

check

$-5(5) = -25$

45. $-2+(-4)=-3n$

$$-6=-3n$$

$$\frac{-6}{-3}=n$$

$$2=n$$

check

$$-6=-3(2)$$

47. $-9h=-3(-3)$

$$-9h=9$$

$$h=\frac{9}{-9}$$

$$h=-1$$

check

$$-9(-1)=9$$

49. $\frac{t}{-3}=-2$

$$t=-3(-2)$$

$$t=6$$

check

$$\frac{6}{-3}=-2$$

51. $0=\frac{y}{8}$

$$0(8)=y$$

$$0=y$$

check

$$0=\frac{0}{8}$$

53. $\dfrac{x}{-2} = -6 + 3$

$\dfrac{x}{-2} = -3$

$x = -2(-3)$

$x = 6$

check

$\dfrac{6}{-2} = -3$

55. $\dfrac{x}{4} = -5 - 8$

$\dfrac{x}{4} = -13$

$x = 4(-13)$

$x = -52$

check

$\dfrac{-52}{4} = -13$

57. $2y + 8 = -6$

$2y + 8 - 8 = -6 - 8$

$2y = -14$

$y = \dfrac{-14}{2}$

$y = -7$

check

$2(-7) + 8 = -6$

59. $-21 = 4h - 5$

$-21 + 5 = 4h - 5 + 5$

$-16 = 4h$

$\dfrac{-16}{4} = h$

$-4 = h$

check

$-21 = 4(-4) - 5$

61. $-3y+1=16$

$$-3y=15$$

$$y=\frac{15}{-3}$$

$$y=-5$$

check

$$-3(-5)+1=16$$

63. $8=-3x+2$

$$6=-3x$$

$$\frac{6}{-3}=\frac{-3x}{-3}$$

$$-2=x$$

65. $-35=5-4x$

$$-40=-4x$$

$$\frac{-40}{-4}=\frac{-4x}{-4}$$

$$10=x$$

67. $4-5x=34$

$$-5x=30$$

$$\frac{-5x}{-5}=\frac{30}{-5}$$

$$x=-6$$

69. $-5-6-5x=4$

$$-11-5x=4$$

$$-5x=15$$

$$\frac{-5x}{-5}=\frac{15}{-5}$$

$$x=-3$$

71. $4-6x=-5-9$

$$4-6x=-14$$

$$-6x=-18$$

$$\frac{-6x}{-6}=\frac{-18}{-6}$$

$$x=3$$

73. $\frac{h}{-6}+4=5$

$\frac{h}{-6}=1$

$h=-6$

75. $-2(4)=\frac{t}{-6}+1$

$-9=\frac{t}{-6}$

$54=t$

77. $0=6+\frac{c}{-5}$

$-6=\frac{c}{-5}$

$30=c$

79. $-1=-8+\frac{h}{-2}$

$7=\frac{h}{-2}$

$-14=h$

81. $2x+3(0)=-6$

$2x=-6$

$x=-3$

83. $2(0)-2y=4$

$-2y=4$

$y=-2$

85. $-x=8$

$x=-8$

Check

$-(-8)=8$

87. $-15=-k$

$15=k$

Check

$-15=-(15)$

Applications

89. SHARKS

Analyze the Problem

- The first observations were at –120 feet.
- The next observations were at –75 feet.
- We must find how many feet the cage was raised.

Form an Equation

- Let x = the number of feet the cage was raised.

Key Word: Raised **Translation**: Add

$-120 + x = -75$

Solve the Equation

$$-120 + x = -75$$

$$-120 + x + 120 = -75 + 120$$

$$x = 45$$

State the Conclusion

The shark cage was raised 45 feet.

Check the Result

If we add the number of feet the cage was raised to the first position, we get $-120 + 45 = -75$. The answer checks.

91. DREDGING A HARBOR

Let x be the amount the harbor is dredged.

$$-47 - x = -65$$

$$-x = -18$$

$$x = 18$$

The harbor was dredged another 18 feet.

93. FOOTBALL

Let r be the rushing total in the second half.

$$43 + r = -8$$

$$43 + r - 43 = -8 - 43$$

$$r = -51$$

The rushing total in the second half was –51 yards.

95. MARKET SHARE

Let m represent market share points.

$$-43 + m = -9$$

$$-43 + m + 43 = -9 + 43$$

$$m = 34$$

The company picked up 34 market share points.

97. PRICE REDUCTION

Let d represent the price drop.

$$\frac{60}{12} = d$$

$$\$5 = d$$

The monthly price drop was $5.

99. ELECTION POLLS

Let s represent the support gained.

$$-31 + s = -2$$

$$-31 + s + 31 = -2 + 31$$

$$s = 29$$

This candidate gained 29 points.

101. INTERNATIONAL TIME ZONES

Let S represent Seattle's time zone.

$$S = 9 - 17$$

$$S = -8$$

Seattle is in zone –8.

Writing

103. Answers may vary

Review

105. $5^6 = 5 \cdot 5 \cdot 5 \cdot 5 \cdot 5 \cdot 5$

107. $7 + 3y = 43$

$$3y = 36$$

$$y = 12$$

109. $16 \div 8 = \dfrac{16}{8}$

Chapter 2 Key Concepts

1. Stocks fell five points, –5.
3. Thirty seconds before going on the air, –30.
5. Ten degrees above normal, +10.
7. Two hundred five dollars overdrawn, –\$205.
9.

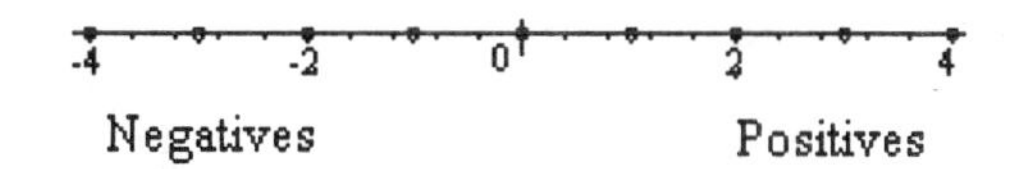

11. $x < y$

13. Addition
    Like Signs – Add the absolute values and maintain the common sign.
    Unlike Signs – Subtract the absolute values and maintain the sign of the number with the largest absolute value.

15. Multiplication
    Like Signs – The product is positive.
    Unlike Signs – The product is negative.

Chapter 2 Review

Section 2.1 An Introduction to the Integers

1. a. $\{-3, -1, 0, 4\}$

   -3 -2 -1 0 1 2 3 4

   b. Integers greater than –3 but less than 4

   -2 -1 0 1 2 3

3. WATER PRESSURE
   A depth of thirty-three feet may be represented as –33.

5. a. $|-4| = 4$
   b. $|0| = 0$
   c. $|-43| = 43$
   d. $-|12| = -12$

Section 2.2 Addition of Integers

7. a. $-(-12) = 12$
   b. $-(8) = -8$
   c. $-(-8) = 8$
   d. $-0 = 0$

9. a. $-6 + (-4) = -10$
   b. $-23 + (-60) = -83$
   c. $-1 + (-4) + (-3) = -8$
   d. $-4 + 3 = -1$
   e. $-28 + 140 = 112$
   f. $9 + (-20) = -11$
   g. $3 + (-2) + (-4) = -3$
   h. $(-2 + 1) + [(-5) + 4] = -1 + (-1) = -2$

11. a. The additive inverse of -11 is 11.
    b. The additive inverse of 4 is -4.

Section 2.3. Subtraction of Integers

13. a. $5-8=-3$
 b. $-9-12=-21$
 c. $-4-(-8)=-4+8=4$
 d. $-6-106=-112$
 e. $-8-(-2)=-8+2=-6$
 f. $7-1=6$
 g. $0-37=-37$
 h. $0-(-30)=0+30=30$

15. a. $-9-7+12=-16+12=-4$
 b. $7-[(-6)-2]=7-(-8)=7+8=15$
 c. $1-(2-7)=1-(-5)=1+5=6$
 d. $-12-(6-10)=-12-(-4)=-12+4=-8$

17. GOLD MINING
 The depth of the second discovery is $150+75=225$ feet down or –225 feet.

19. RECORD TEMPERATURES
 Alaska $100-(-80)=100+80=180$ degrees
 Virginia $110-(-30)=110+30=140$ degrees

Section 2.4 Multiplication of Integers

21. a. $(-6)(-2)(-3)=12(-3)=-36$
 b. $4(-3)3=-12(3)=-36$
 c. $0(-7)=0$
 d. $(-1)(-1)(-1)(-1)=1(-1)(-1)=-1(-1)=1$

23. a. $(-5)^2=25$
 b. $(-2)^5=-32$
 c. $(-8)^2=64$
 d. $(-4)^3=-64$

25. The difference between -2^2 and $(-2)^2$ is the base, the first expression has two as a base and the second expression has negative two as a base.
 $-2^2=-4$ and $(-2)^2=4$

Section 2.5 Division of Integers

27. a. $\dfrac{-14}{7} = -2$

b. $\dfrac{25}{-5} = -5$

c. $-64 \div 8 = -8$

d. $\dfrac{-202}{-2} = 101$

29. PRODUCTION TIME

$\dfrac{12 \text{ minutes}}{6 \text{ months}} = 2$ minutes per month or -2

Section 2.6 Order of Operations and Estimation

31. a. $-4\left(\dfrac{15}{-3}\right) - 2^3 = -4(-5) - 8 = 20 - 8 = 12$

b. $-20 + 2(12 - 5 \cdot 2) = -20 + 2(2) = -20 + 4 = -16$

c. $-20 + 2[12 - (-7 + 5)^2] = -20 + 2[12 - (-2)^2]$

$= -20 + 2[12 - 4]$

$= -20 + 2[8]$

$= -20 + 16$

$= -4$

d. $8 - |-3 \cdot 4 + 5| = 8 - |-12 + 5|$

$= 8 - |-7|$

$= 8 - 7$

$= 1$

33. a. $-89 + 57 + (-42) \approx -70$

b. $\dfrac{-507}{-24} \approx 20$

c. $(-681)(9) \approx -7{,}000$

d. $317 - (-775) \approx 1{,}100$

Section 2.7 Solving Equations Involving Integers

35. a. $t+(-8)=-18$

$$t=-18+8$$
$$t=-10$$

b. $\dfrac{x}{-3}=-4$

$$x=-3(-4)$$
$$x=12$$

c. $y+8=0$

$$y=-8$$

d. $-7m=-28$

$$m=\frac{-28}{-7}$$
$$m=4$$

37. a. $-5t+1=-14$

$$-5t=-15$$
$$t=3$$

Check

$$-5(3)+1=-15+1=-14$$

b. $3(2)=2-2x$

$$6-2=-2x$$
$$4=-2x$$
$$-2=x$$

Check

$$6=2-2(-2)$$

c. $\dfrac{x}{-4} - 5 = -1 - 1$

$\dfrac{x}{-4} = 3$

$x = -4(3)$

$x = -12$

Check

$\dfrac{-12}{-4} - 5 = -2$

d. $c - (-5) = 5$

$c + 5 = 5$

$c = 0$

Check

$0 - (-5) = 5$

39. CREDIT CARD PROMOTION

Let p = the number of customers

$8p = 968$

$p = 121$

121 customers applied for credit.

Chapter 2 Test

1. a. $-8 > -9$
   b. $-8 < |-8|$ since $|-8| = 8$
   c. The opposite of $5 < 0$ since the opposite of 5 is – 5.

3. SCHOOL ENROLLMENT
   Monroe will face the greatest shortage, -2,488.

5. a. $-65 + 31 = -34$
   b. $-17 + (-17) = -34$
   c. $[6 + (-4)] + [-6 + (-4)] = 2 + (-10) = -8$

7. a. $-10 \cdot 7 = -70$
   b. $-4(-2)(-6) = 8(-6) = -48$
   c. $(-2)(-2)(-2)(-2) = 4(4) = 16$
   d. $-55(0) = 0$

9. a. $\frac{-32}{4} = -8$

   b. $\frac{8}{6-6} = \frac{8}{0}$ is undefined

   c. $\frac{-5}{1} = -5$

   d. $\frac{0}{-6} = 0$

11. GEOGRAPHY
    $436 - 282 = 154$ feet difference in elevation.

13. a. $(-4)^2 = 16$
    b. $-4^2 = -16$
    c. $(-4-3)^2 = (-7)^2 = 49$

15. $4 - (-3)^2 + 6 = 4 - 9 + 6 = 1$

17. $-10+2[6-(-2)^2(-5)] = -10+2[6-4(-5)]$
$$= -10+2[6+20]$$
$$= -10+2(26)$$
$$= 42$$

19. $c-(-7) = -8$
$$c+7 = -8$$
$$c = -15$$

21. $\frac{x}{-4} = 10$
$$x = -40$$

23. $-5 = -6a+7$
$$-12 = -6a$$
$$2 = a$$
Check
$$-5 = -6(2)+7$$

25. CHECKING ACCOUNT
$$b+225 = -19$$
$$b = -\$244$$
Before the deposit the balance was -$244.

27. $5(-4) = (-4)+(-4)+(-4)+(-4)+(-4) = -20$

Chapter 1 – 2 Cumulative Review Exercises

1. The natural numbers are 1, 2, 5, 9.

3. The negative numbers are –2, –1.

5. 6 is in the thousands column.

7. To the nearest hundred 7,326,500.

9. BIDS
   If the award goes to the lowest bid then the contract should be given to CRF Cable.

11. $237 + 549 = 786$

13. $5{,}369 - 685 = 4{,}684$

15. The perimeter is $2(17) + 2(35) = 34 + 70 = 104$ feet.
    The area is $35(17) = 595 \text{ ft}^2$.

17. $435 \cdot 27 = 11{,}745$

19. $4{,}587 \times 67 = 307{,}329$

21. SHIPPING
    In 12 gross there would be $12 \cdot 12 \cdot 12 = 1{,}728$ tennis balls.

23. 17 is prime and odd.

25. 0 is even.

27. $504 = 3(168) = 3(3)(56) = 3^2(2)(28) = 3^2(2)(2)(14) = 3^2 \cdot 2^2 \cdot (2)(7) = 2^3 \cdot 3^2 \cdot 7$

29. $5^2 \cdot 7 = 25(7) = 175$

31. $25 + 5 \cdot 5 = 25 + 25 = 50$

33. SPEED CHECK
    The average speed was $\dfrac{38 + 42 + 36 + 38 + 48 + 44}{6} = \dfrac{246}{6} = 41$ mph.
    The drivers were not obeying the speed limit.

35. $50 = x + 37$
    $13 = x$

37. $5p = 135$
    $p = 27$

39. $\{-2,-1,0,2\}$

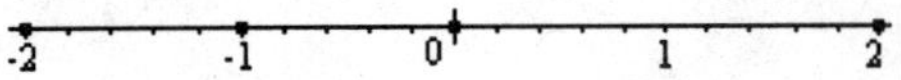

41. $-17<-16$ is a true statement.

43. $-2+(-3)=-5$

45. $-3-5=-8$

47. $(-8)(-3)=24$

49. $\dfrac{-14}{-7}=2$

51. $5+(-3)(-7)=5+21=26$

53. $\dfrac{10-(-5)}{1-2\cdot 3}=\dfrac{10+5}{1-6}=\dfrac{15}{-5}=-3$

55.
$$\begin{aligned} -5t+1&=-14 \\ -5t&=-15 \\ t&=3 \end{aligned}$$

Check $-5(3)+1=-14$

57. BUYING A BUSINESS

Each person's share of the debt was $\dfrac{1{,}512{,}444}{12}=\$126{,}037$.

Section 3.1 Variables and Algebraic Expressions

Vocabulary

1. An algebraic **expression** is a combination of variables, numbers, and the operation symbols for addition, subtraction, multiplication and division.
3. A **variable** is a letter that is used to stand for a number.

Concepts

5. Answers will vary, $10+3x; \frac{x-10}{3}$

7. Mr. Lamb lives further away by 15 miles.

9. In 1999 the profits were 2p and in 2000 the profits were 3p.

11.

Wind Conditions	*Speed of Jet*
In Still Air	500
With the Tail Wind	$500+x$
Against the Head Wind	$500-x$

13. This student should study $\frac{h}{4}$ hours per day.

15. There are $450-x$ inches of tape left.

Notation

17. $x \cdot 8 = 8x$

19. $10 \div g = \frac{10}{g}$

Practice

21. $x-9$

23. $\frac{2}{3}p$

25. $r+6$

27. $d-15$

29. $1-s$

31. $2p$

33. $s+14$

35. $\frac{35}{b}$

37. $x-2$

39. c increased by 7

41. 7 less than c

43. a. There are $60m$ seconds in m minutes.
b. In h hours there are $60 \cdot 60h = 3{,}600h$ seconds.

45. a. Per month the salary is $\frac{s}{12}$.

b. Per week the salary is $\frac{s}{52}$.

47. a. This rope is $12f$ inches long.

b. This rope is $\frac{f}{3}$ yards long.

49. The decibel reading for the concert was $j-5$.

51. There would be $6s$ individual cans.

53. The pad of paper will last $\frac{p}{15}$ days.

55. There was $t+2$ tons of paper collected.

57. Let w represent the width then the length is $w+6$.

59. Let g represent the quantity drained then the remaining amount is $6-g$.

61. $3x+5$

63. $10a+12$

Applications

65. PRESIDENTIAL ELECTIONS
Let N represent the number of votes Nixon received then Kennedy received $N+118{,}550$ votes.

67. THE BEATLES
Let s represent the number of copies of *I Want to Hold Your Hand* then the number of copies of *Hey Jude* sold was $s-2{,}000{,}000$.

Writing

69. Answers may vary.
71. Answers may vary.

Review

73. $-5+(-6)+1=-11+1=-10$

75. $-x=4$

$x=-4$

77. The set of integers are $\{\ldots-3,-2,-1,0,1,2,3\ldots\}$.

79. $-3+(-2)+7=-5+7=2$

Section 3.2 Evaluating Algebraic Expressions and Formulas

Vocabulary

1. A **formula** is an equation that states a known relationship between two or more variables.
3. To evaluate an algebraic expression, we **substitute** specific numbers for the variables in the expression and apply the rules for order of operations.

Concepts

5. By not using the parenthesis around negative eight the expression appears to become a subtraction problem.

7. a. Let x be the length of Part 1 then Part 2 has length $x-40$ inches and Part 3 is $x+16$..
   b. If Part 1 is 60 inches, then Part 2 is $60-40=20$ inches and Part 3 is $60+16=76$ inches.

9. a.

Price	Service	Cost
20	2	22
25	2	27
p	2	$p+2$

   b. The total cost is $T=p+2$.

11. $d=rt$

	Rate	Time	Distance
Bike	12	4	48
Walking	3	t	$3t$
Car	x	3	$3x$

13. a. A health club instructor might use a target heart rate after a workout.
   b. A mechanic might use gas mileage of a car.
   c. A paleontologist might use the age of a fossil.
   d. A realtor might use equity in a home.
   e. A doctor might use a dosage to administer.
   f. An economist might use a cost-of-living index.

Notation

15. a. $d=rt$
   b. $C=\dfrac{5(F-32)}{9}$
   c. $d=16t^2$

Practice

17. $3(4)+5=17$

19. $-(-4)=4$

21. $-4(-10)=40$

23. $\frac{-4-8}{2}=\frac{-12}{2}=-6$

25. $2(-12+9)=2(-3)=-6$

27. $(-5)^2-(-5)-7=25+5-7=23$

29. $8(-2)-(-2)^3=-16-(-8)=-8$

31. $4(5)^2=100$

33. $3(-4)-(-4)^2=-12-16=-28$

35. $\frac{24+3}{3(3)}=\frac{27}{9}=3$

37. $|6-50|=|-44|=44$

39. $-2|-7|-7=-14-7=-21$

41. $\frac{30}{-10}=-3$

43. $-(-1)-8=1-8=-7$

45. $-2(5(2)-1)=-2(9)=-18$

47. $(-3)^2-4(4)(-1)=9+16=25$

49. $5^2-(-2)^2=25-4=21$

51. $\frac{50-6(5)}{-4}=\frac{20}{-4}=-5$

53. $-5(-2)(-1)(3)+1=-30+1=-29$

55. $5(-3)^2(-1)=5(9)(-1)=-45$

57. $|(-2)^2-(-5)^2|=|4-25|=21$

59. The price of a snow cone is $20+50=70$ cents.

61. The profit was $\$13{,}500-\$5{,}300=\$8{,}200$.

63. The retail price of the bracelet is $\$18+\$5=\$23$.

65. The distance covered was $d=60(5)=300$ miles.

67. The temperature is $C=\dfrac{5(14-32)}{9}=-10$ degrees Celsius.

69. The average score is $\dfrac{254+225+238}{3}=239$.

71. The ball has fallen $d=16(2)^2=64$ feet.

Applications

73. FINANCIAL STATEMENT

	Dec. '98	Dec. '97	Dec. '96
Revenue (r)	5,213	5,079	4,814
Cost of Goods (c)	2,053	2,051	1,921
Gross Profit $p = r\text{-}c$	3,160	3,028	2,893

75. THERMOMETER SCALE

Starting at the top of the scale and working down using $C=\dfrac{5(F-32)}{9}$,

Fahrenheit	Celsius
86	30
59	15
23	-5

77. FALLING OBJECT

$d=16t^2$

Time Falling	Distance Traveled	Time Intervals
1 sec	$16(1)^2=16$	0 to 1 sec $16-0=16$ ft
2 sec	$16(2)^2=64$	1 to 2 sec $64-16=48$ ft
3 sec	$16(3)^2=144$	2 to 3 sec $144-64=80$ ft
4 sec	$16(4)^2=256$	3 to 4 sec $256-144=112$ ft

79. CUSTOMER SATISFACTORY SURVEY

There were 53 marks of 5, 26 marks of 3, and 9 marks of 1, so the average was

$$\frac{53(5)+26(3)+9(1)}{53+26+9}=\frac{352}{88}=4.$$

Writing

81. Answers may vary.
83. Answers may vary.
85. Answers may vary.

Review

87. 17, 37, and 41 are prime numbers.
89. $|-2+(-5)|=|-7|=7$
91. Division by 3 is performed on the variable.
93. $-3-(-6)=-3+6=3$

Section 3.3 Simplifying Algebraic Expressions and the Distributive Property

Vocabulary

1. The **distributive** property tells us how to multiply $5(x+7)$. After doing the multiplication to obtain $5x+35$, we say that the parentheses have been **removed**.
3. When an algebraic expression is simplified, the result is an **equivalent** expression.

Concepts

5. $x(y+z)=xy+xz$
7. $(w+7)5$ would be termed the right distributive property.
9. Two groups of 6 plus three groups of 6 is 5 groups of 6. Therefore, $6{\bullet}2+6{\bullet}3=6(2+3)$
11. $-(y+9)=-y-9$

Notation

13. $-5(7n)=(-5{\bullet}7)n$
$$=-35n$$

15. $-9(-4-5y)=(-9)(-4)-(-9)(5y)$
$$=36-(-45y)$$
$$=36+45y$$

17. a. $-(-x)=x$
 b. $x-(-5)=x+5$
 c. $5x-10y+(-15)=5x-10y-15$
 d. $5{\bullet}x=5x$

Practice

19. $2(6x)=12x$
21. $-5(6y)=-30y$
23. $-10(-10t)=100t$
25. $(4s)3=12s$
27. $2c{\bullet}7=14c$
29. $-5{\bullet}8h=-40h$
31. $-7x(6y)=-42xy$
33. $4r{\bullet}4s=16rs$
35. $2x(5y)(3)=30xy$
37. $5r(2)(-3b)=-30rb$

39. $5 \cdot 8c \cdot 2 = 80c$
41. $(-1)(-2e)(-4) = -8e$
43. $4(x+1) = 4x+4$
45. $4(4-x) = 16-4x$
47. $-2(3e+3) = -6e-6$
49. $-8(2q-6) = -16q+48$
51. $-4(-3-5s) = 12+20s$
53. $(7+4d)6 = 42+24d$
55. $(5r-6)(-5) = -25r+30$
57. $(-4-3d)6 = -24-18d$
59. $3(3x-7y+2) = 9x-21y+6$
61. $-3(-3z-3x-5y) = 9z+9x+15y$
63. $-(x+3) = -x-3$
65. $-(4t+5) = -4t-5$
67. $-(-3w-4) = 3w+4$
69. $-(5x-4y+1) = -5x+4y-1$
71. $2(4x)+2(5) = 2(4x+5)$
73. $-4(5)-3x(5) = (-4-3x)5$
75. $-3(4y)-(-3)(2) = -3(4y-2)$
77. $3(4)-3(7t)-3(5s) = 3(4-7t-5s)$

Writing

79. Answers may vary
81. Answers may vary

Review

83. $|-6+1| = |-5| = 5$
85. Product is associated with multiplication.
    Quotient is associated with division.
    Difference is associated with subtraction.
    Sum is associated with addition.
87. $-6 > -7$
89. Carpeting a room and painting a room involve area.

Section 3.4 Combining Like Terms

Vocabulary

1. A **term** is a number or a product of a number and one or more variables.
3. The **perimeter** of a geometric figure is the distance around it.
5. $2(x+3)=2x+2(3)$ is an example of the **distributive** property.
7. Simplifying the sum (or difference) of like terms is called **combining** like terms.

Concepts

9. a. x is used as a term.
   b. x is used as a factor.
   c. x is used as a factor.
   d. x is used as a factor.

11. a. 11 is the numerical coefficient.
    b. 8 is the numerical coefficient.
    c. –4 is the numerical coefficient.
    d. 1 is the numerical coefficient.
    e. –1 is the numerical coefficient.
    f. 102 is the numerical coefficient.

13.

Term	Coefficient	Variable Part
$6m$	6	m
$-75t$	–75	T
w	1	W
$4bh$	4	Bh

15. Underlining helps identify like terms.
17. The total distance is $(d+15)+d=2d+15$ miles.
19. The diagram illustrates that when summing like terms add the coefficients and keep the variable.

Notation

21. $5x+7x=(5+7)x=12x$
23. $2(x-1)+3x=2x-2+3x=5x-2$
25. a. P represents perimeter of a rectangle.
    b. $2l$ represents the product of two and the rectangle length.
    c. $2w$ represents the product of two and the rectangle width.

Practice

27. The terms are $3x^2, -5x,$ and 4.
29. The terms are $5, 5t, -8t$ and 4.
31. To make these like terms there must be an exponent of 2.
33. To make these like terms there must be an exponent of 5.
35. $6t + 9t = 15t$
37. $5s - s = 4s$
39. $-5x + 6x = x$
41. $-5d + 9d = 4d$
43. $3e - 7e = -4e$
45. $h - 7$ cannot be simplified.
47. $4z - 10z = -6z$
49. $-3x - 4x = -7x$
51. $2t - 2t = 0$
53. $-6s + 6s = 0$
55. $x + x + x + x = 4x$
57. $2x + 2y$ cannot be simplified.
59. $0 - 2y = -2y$
61. $3a - 0 = 3a$
63. $6t + 9 + 5t + 3 = 11t + 12$
65. $3w - 4 - w - 1 = 2w - 5$
67. $-4r + 8R + 2R - 3r + R = -7r + 11R$
69. $-45d - 12a - 5d + 12a = -50d$
71. $4x - 3y - 7 + 4x - 2 - y = 8x - 4y - 9$
73. $$\begin{aligned} 4(x+1) + 5(6+x) &= 4x + 4 + 30 + 5x \\ &= 9x + 34 \end{aligned}$$
75. $$\begin{aligned} 5(3-2s) + 4(2-3s) &= 15 - 10s + 8 - 12s \\ &= -22s + 23 \end{aligned}$$
77. $$\begin{aligned} -4(6-4e) + 3(e+1) &= -24 + 16e + 3e + 3 \\ &= 19e - 21 \end{aligned}$$
79. $3t - (t-8) = 3t - t + 8 = 2t + 8$
81. $$\begin{aligned} -2(2-3x) - 3(x-4) &= -4 + 6x - 3x + 12 \\ &= 3x + 8 \end{aligned}$$
83. $$\begin{aligned} -4(-4y+5) - 6(y+2) &= 16y - 20 - 6y - 12 \\ &= 10y - 32 \end{aligned}$$

Applications

85. MOBILE HOME DESIGN

There are four faces of the mobile home to consider, but remember that the front and back will use the same amount, as will the two ends. For the front and back they will need $2(10)+2(60)=140$ feet each, or 280 feet. For the side they will need $4(10)=40$ feet, or 80 feet for both sides. They need a total of $80+280=360$ feet of trim at a cost of $360(0.80)=\$288$.

87. PARTY PREPARATIONS

Slow	Fast	Floor Size	Perimeter
8	5	9×9	$4(9)=36$ feet
14	9	12×12	$4(12)=48$ feet
22	15	15×15	$4(15)=60$ feet
32	20	18×18	$4(18)=72$ feet
50	30	21×21	$4(21)=84$ feet

Writing

89. Answers may vary.
91. Answers may vary.

Review

93. $-4t-3=-11$

$$-4t=-8$$

$$t=2$$

95. $100=2^2\cdot 5^2$

97. The **absolute value** of a number is the distance between it and 0 on the number line.

Section 3.5 Simplifying Expressions to Solve Equations

Vocabulary

1. To **solve** an equation means to find all values of the variable that make the equation a true statement.
3. In $2(x+4)$, to remove parentheses means to apply the **distributive** property.
5. The phrase "**combine** like terms" refers to the operations of addition and subtraction.

Concepts

7. When negative five is substituted a false statement is made.

$$5x - 3x = -9$$
$$5(-5) - 3(-5) = -10 \neq -9$$

9. We should subtract $5k$ from both sides.

11. a. $4x$ should be subtracted from both sides.
    b. $2x$ should be subtracted from both sides.

13. a. $3t - t - 8 = 2t - 8$
    b. $3t - t = -8$
    $$2t = -8$$
    $$t = -4$$
    c. $3t - t - 8$
    $$3(-4) - (-4) - 8 = -12 + 4 - 8 = -16$$

Notation

15. $4x - 2x = -20$

$$2x = -20$$
$$\frac{2x}{2} = \frac{-20}{2}$$
$$x = -10$$

17. $$\begin{aligned} 5(x-9) &= 5 \\ 5x-5(9) &= 5 \\ 5x-45 &= 5 \\ 5x-45+45 &= 5+45 \\ 5x &= 50 \\ \frac{5x}{5} &= \frac{50}{5} \\ x &= 10 \end{aligned}$$

Practice

19. $5(3)+8 \stackrel{?}{=} 4(3)+11$

$15+8 \stackrel{?}{=} 12+11$

$23=23$

21. $2(12-1) \stackrel{?}{=} 33$

$2(11) \stackrel{?}{=} 33$

$22 \neq 33$

23. $$\begin{aligned} 3x+6x &= 54 \\ 9x &= 54 \\ x &= 6 \end{aligned}$$

25. $$\begin{aligned} 6x-3x &= 9 \\ 3x &= 9 \\ x &= 3 \end{aligned}$$

27. $$\begin{aligned} 60 &= 3v-5v \\ 60 &= -2v \\ -30 &= v \end{aligned}$$

29. $$\begin{aligned} -28 &= -m+2m \\ -28 &= m \end{aligned}$$

31. $$\begin{aligned} x+x+6 &= 90 \\ 2x &= 84 \\ x &= 42 \end{aligned}$$

33. $T + T - 17 = 57$

$2T = 74$

$T = 37$

35. $600 = m - 12 + m$

$612 = 2m$

$306 = m$

37. $1{,}500 = b + 30 + b$

$1{,}470 = 2b$

$735 = b$

39. $7x = 3x + 8$

$4x = 8$

$x = 2$

41. $x - 14 = 2x$

$-14 = x$

43. $9t - 40 = 14t$

$-40 = 5t$

$-8 = t$

45. $25 + 4j = 9j$

$25 = 5j$

$5 = j$

47. $-48 + 12t = 16t$

$-48 = 4t$

$-12 = t$

49. $-5g - 40 = -15g$

$10g - 40 = 0$

$10g = 40$

$g = 4$

51. $3s + 1 = 4s - 7$

$1 = s - 7$

$8 = s$

53. $50a-1=60a-101$
$$-1=10a-101$$
$$100=10a$$
$$10=a$$

55. $-7+5r=83-10r$
$$-7+15r=83$$
$$15r=90$$
$$r=6$$

57. $100-y=100+y$
$$100=100+2y$$
$$0=2y$$
$$0=y$$

59. $2(x+6)=4$
$$2x+12=4$$
$$2x=-8$$
$$x=-4$$

61. $-16=2(t+2)$
$$-16=2t+4$$
$$-20=2t$$
$$-10=t$$

63. $-3(2w-3)=9$
$$-6w+9=9$$
$$-6w=0$$
$$w=0$$

65. $-(c-4)=3$
$$-c+4=3$$
$$-c=-1$$
$$c=1$$

67. $4(p-2)=0$
$$4p-8=0$$
$$4p=8$$
$$p=2$$

69. $2(4y+8)=3(2y-2)$
$$8y+16=6y-6$$
$$2y+16=-6$$
$$2y=-22$$
$$y=-11$$

71. $16-(x+3)=-13$
$$16-x-3=-13$$
$$13-x=-13$$
$$-x=-26$$
$$x=26$$

73. $5-(7-y)=-5$
$$5-7+y=-5$$
$$-2+y=-5$$
$$y=-3$$

75. $2x+3(x-4)=23$
$$2x+3x-12=23$$
$$5x-12=23$$
$$5x=35$$
$$x=7$$

77. $10q+3(q-7)=18$
$$10q+3q-21=18$$
$$13q-21=18$$
$$13q=39$$
$$q=3$$

Writing

79. Answers may vary
81. Answers may vary

Review

83. $-7-9=-16$

85. $\dfrac{-8+2}{-2+4}=\dfrac{-6}{2}=-3$

87. $-(-5)=5$

89. The sign of the product of two negative integers is positive.

Section 3.6 Problem Solving

Vocabulary

1. The words *increased by*, *longer*, *taller*, *higher*, *total*, and *more than* indicate the operation of **addition** should be used.

Concepts

3. BUSINESS ACCOUNTS
   In x months the salesman will add $5x$ accounts.

5. SERVICE STATION
   The smaller tank holds $g-100$ gallons.

7. OCEAN TRAVEL
   The passenger ship traveled $3m$ miles.

9. $l = 2w$

11. COMMISSION

Type	Sold	Commission	Total
Dress	10	3	30
Athletic	12	2	24
Child's	x	5	$5x$
Sandal	$9-x$	4	$4(9-x)$

13. a. There are 9 pairs of shoes stored in the rack.
    b. There would be $9-d$ pairs of athletic shoes.

15. AIRLINE SEATING

Analyze the problem

- There are 88 seats on the plane.
- There are 10 times as many economy as first-class seats.
- Find the number of first-class seats.

Form an equation

Since the number of economy seats is related to the number of first-class seats, we let $x =$ the number of first-class seats.

Key Phrase: ten times as many

Translation: multiply by 10

So $10x =$ the number of economy seats.

The number of first-class seats plus the number of economy seats is 88.

$x+10x=88$

Solve the equation

$x+10x=88$

$11x=88$

$x=8$

State the conclusion: There are 8 first-class seats.

Check the result: If there are 8 first-class seats, there are $10 \bullet 8 = 80$ economy seats. Adding 8 and 80, we get 88. The answer checks.

17. BUSINESS ACCOUNTS

Let x represent months

$15+5x=100$

$5x=85$

$x=17$

It will take him 17 months to reach 100 accounts.

19. ANTIQUE COLLECTING

Let y represent years

$56+4y=100$

$4y=44$

$y=11$

In 11 years she will have 100 spoons.

21. APARTMENT RENTAL

Let r represent rent

$100+r=425$

$r=\$325$

The total rent is $3(325)=\$975$.

23. BOTTLED WATER DELIVERY

Let b represent buildings

$300-3b=117$

$-3b=-183$

$b=61$

This driver delivered to 61 buildings.

25. SERVICE STATION

Let g represent the amount held in the premium tank, then in the regular tank there is $g-100$ gallons.

$g+(g-100)=700$

$2g=800$

$g=400$

There are 400 gallons in the premium tank.

27. OCEAN TRAVEL

The total distance traveled was 84 miles; let f represent the distance traveled by the freighter.

$f+3f=84$

$4f=84$

$f=21$

The freighter was 21 miles from port.

29. INTERIOR DECORATING

$l=2w$ and $P=60$, then using .

$P=2l+2w$

$P=2(2w)+2w=6w$

$60=6w$

$10=w$

The width of the room is 10 feet.

31. COMMERCIALS

Let t represent the amount of time dedicated to commercials.

$t+(t+18)=30$

$2t=12$

$t=6$

There were 6 minutes of commercials.

33. COMMISSION

Shoe Type	Commission	Total Sold	Total Commission
Dress	3	d	$3d$
Athletic	2	$9-d$	$2(9-d)$

$$3d+2(9-d)=24$$
$$3d+18-2d=24$$
$$d=6$$

The salesman sold 6 pairs of dress shoes and $9-6=3$ pairs of athletic shoes.

35. MOVER'S PAY SCALE

Mover	Part-Time	Hours	Total
Rate	6	h	$6h$
Stair Rate	9	$20-h$	$9(20-h)$

$$6h+9(20-h)=138$$
$$6h+180-9h=138$$
$$-3h=-42$$
$$h=14$$

This mover worked 14 hours for \$6 per hour and 20 – 14 = 6 hours for \$9 per hour.

Writing

37. Answers may vary.
39. Answers may vary.

Review

41. This illustrates the associative property of addition.

43. $-10^2=-100$

45. Subtraction of a number is the same as **addition** of the opposite of that number.

47. $2\cdot2\cdot2\cdot5\cdot5=2^3\cdot5^2$

Chapter 3 Key Concept

1. Do all calculations within **parentheses** and other grouping symbols, following the order listed in Steps 2 – 4 below and working from the **innermost** pair to the **outermost** pair.

3. Do all **multiplications** and **divisions** as they occur from left to right.

5. $-10+4-3^2=-10+4-9=-6-9=-15$

7. $-2(-3)-12\div 6\bullet 3=6-2\bullet 3=0$

9. $2(4+3\bullet 2)^2-(-6)=2(10)^2+6=206$

11. $1-2|8(-2)-(-2)^3|=1-2|-16-(-8)|$

$$=1-2|-8|$$
$$=1-2(8)$$
$$=-15$$

13. $P=2(30)+2(16)$

$P=60+32$

$P=92$ feet

15. $(3x)4-2(5x)+x=12x-10x+x=3x$

17. $15-3(-3)\stackrel{?}{=}23$

$15+9\stackrel{?}{=}23$

$24\neq 23$

19. To isolate x on the left hand side start by undoing the subtraction first then undo the multiplication.

Chapter 3 Review

Section 3.1 Variables and Algebraic Expressions

1. a. Brandon is closer by 250 miles.
   b. The height of the ceiling is $h+7$ feet.

3. a. There will be $\frac{c}{6}$ children in each room.
   b. The new price is $\$1,000-\$x$.

5. a. There are $12x$ eggs in x dozen.
   b. There are $\frac{d}{7}$ days.

Section 3.2 Evaluating Algebraic Expressions and Formulas

7. a. $-2(-3)+6=6+6=12$
   b. $\frac{6-(-2)}{1+(-2)}=\frac{8}{-1}=-8$
   c. $6^2-4(4)(-4)=36+64=100$
   d. $\frac{-2(-2)^3}{1-2-3}=\frac{-2(-8)}{-4}=\frac{16}{-4}=-4$

9. SALE PRICE
   The sale price will be $315-37=\$278$.

11. ANNUAL PROFIT
    a. 1998 had the most revenue.
    b. The largest profit was in 2000.
    c. Costs decreased over this three-year period.

13. DISTANCE FALLEN
    Using $d=16t^2$, in three seconds the hammer fell
    $d=16(3)^2$
    $d=144$ feet.

Section 3.3 Simplifying Algebraic Expressions and the Distributive Property

15. a. $-2(5x) = -10x$
    b. $-7x(-6y) = 42xy$
    c. $4d \bullet 3e \bullet 5 = 60de$
    d. $(4s)8 = 32s$
    e. $-1(-e)(2) = 2e$
    f. $7x \bullet 7y = 49xy$
    g. $4 \bullet 3k \bullet 7 = 84k$
    h. $(-10t)(-10) = 100t$

17. a. $-(6t-4) = -6t+4$
    b. $-(5+x) = -5-x$
    c. $-(6t-3s+1) = -6t+3s-1$
    d. $-(-5a-3) = 5a+3$

Section 3.4 Combining Like Terms

19. a. Factor
    b. Term
    c. Factor
    d. Term

21. a. $3x+4x = 7x$
    b. $-3t-6t = -9t$
    c. $2z+(-5z) = -3z$
    d. $6x-x = 5x$
    e. $-6y-7y-(-y) = -12y$
    f. $5w-8-4w+3 = w-5$
    g. $-45d-2a+4a-d = -46d+2a$
    h. $5y+8h-3+7h+5y+2 = 10y+15h-1$

23. HOLIDAY LIGHTS
    The perimeter of the house is $2(35)+2(42) = 154$ feet. The perimeter of the two windows is $2(4)(5) = 40$ feet. 154 + 40 = 194 feet of lights are needed to decorate this house.

Section 3.5 Simplifying Expressions to Solve Equations

25. a. $5a - 3a = -36$

$$2a = -36$$
$$a = -18$$

b. $3x - 4x = -8$

$$-x = -8$$
$$x = 8$$

c. $7x = 3x - 12$

$$4x = -12$$
$$x = -3$$

d. $5(y - 15) = 0$

$$5y - 75 = 0$$
$$5y = 75$$
$$y = 15$$

e. $3a - (2a - 1) = -2$

$$3a - 2a + 1 = -2$$
$$a = -3$$

f. $15 = 5b + 1 + 2b$

$$14 = 7b$$
$$2 = b$$

g. $-6(2x + 3) = -(5x - 3)$

$$-12x - 18 = -5x + 3$$
$$-21 = 7x$$
$$-3 = x$$

h. $-3(2x + 4) - 4 = -40$

$$-6x - 12 = -36$$
$$-6x = -24$$
$$x = 4$$

Section 3.6 Problem Solving

27. REFRESHMENTS

There are $56-c$ cups of coffee left.

29. FITNESS

If she bikes b miles, then she jogs $b-8$ miles and the sum of these is 18 miles.

$$b+(b-8)=18$$
$$2b=26$$
$$b=13$$

She bicycles 13 miles.

31. HEALTH FOOD

Let d represent the number of \$3 drinks sold, then there were $50-d$ \$4 drinks sold.

$$3d+4(50-d)=185$$
$$3d+200-4d=185$$
$$-d=-15$$
$$d=15$$

There were 15 \$3 drinks sold and 50 – 15 = 35 \$4 drinks sold.

Chapter 3 Test

1. a. $r-2$
   b. $3xy$

3. a. $\frac{4-16}{4}=\frac{-12}{4}=-3$
   b. $2(-2)^2-3(-2-4)=2(4)-3(-6)$
   $$=8+18$$
   $$=26$$

5. PROFIT
   Using $P=R-C$ where revenue is 40,000 + 15,000 = \$55,000 and costs are 13,000 + 5,000 = \$18,000 we get
   $P=55,000-18,000$
   $P=\$37,000$

7. METER READING
   The average meter reading was
   $$m=\frac{4+2+(-4)+(-4)+4+2+(-4)+2+4+4}{10}$$
   $$m=\frac{10}{10}$$
   $$m=1$$

9. AIR CONDITIONING
   Using $C=\frac{5(F-32)}{9}$
   $$C=\frac{5(59-32)}{9}$$
   $$C=\frac{5(27)}{9}$$
   $$C=15^0$$

11. a. x is used as a factor.
    b. x is used as a term.

13. a. The three terms are $8x^2, -4x$, and -6.
    b. The coefficient of the first term is 8.

15. $4(y+3)-5(2y+3)=4y+12-10y-15=-6y-3$

17. $6r = 2r - 12$

$4r = -12$

$r = -3$

19. $6 - (y - 3) = 19$

$6 - y + 3 = 19$

$9 - y = 19$

$-y = 10$

$y = -10$

21. DRIVING SCHOOL

From the instructions, students spend 14 – 2 = 12 hours in class Monday through Thursday.

Therefore there are $\frac{12}{4} = 3$ hours per classroom session.

23. Like terms have exactly the same variables and exponents.

25. No; We simplify expressions and solve equations.

Chapters 1 – 3 Cumulative Review Exercises

1. GASOLINE
   358,600,000 gallons

3. 54,604

5. 23,115

7. $683 + 459 = 1{,}142$

9. $4 \cdot 5 = 5 + 5 + 5 + 5 = 20$

11. a. $F_{18} = \{1, 2, 3, 6, 9, 18\}$
    b. The prime factorization of 18 is $3^2 \cdot 2$

13. 27 has factors other than one and itself, thus it is not prime. For example, $3 \cdot 9 = 27$.

15. $250 = \dfrac{y}{2}$
    $500 = y$

17. Graph the integers greater than –3 and less than 4.

-4 -3 -2 -1 0 1 2 3 4 5

19. This is a false statement.

21. $-25 + 5 = -20$

23. $-25(5)(-1) = 125$

25. $\dfrac{(-6)^2 - 1^5}{-4 - 3} = \dfrac{36 - 1}{-7} = \dfrac{35}{-7} = -5$

27. $-3^2 = -(3 \cdot 3) = -9$
    $(-3)^2 = (-3)(-3) = 9$

29. $-4x + 4 = -24$
    $-4x = -28$
    $x = 7$

31. Division by zero is undefined, $\dfrac{5}{0}$; Division of zero, $\dfrac{0}{5}$.

33. "h increased by 12" is $h + 12$.

35. $x^2 = x \cdot x$

$2x = x + x$

37. $d = rt$

	Rate	Time	Distance
Truck	55	4	$55(4) = 220$

39. $-6(-4t) = 24t$

41. As a term, $x + 24$, as a factor $25x$.

43. $5b + 8 - 6b - 7 = -b + 1$

45. $7 + 2x = 2 - (4x + 7)$

$7 + 2x = 2 - 4x - 7$

$7 + 2x = -4x - 5$

$12 = -6x$

$-2 = x$

Section 4.1 The Fundamental Property of Fractions

Vocabulary

1. For the fraction $\frac{7}{8}$, 7 is the **numerator** and 8 is the **denominator**.

3. A **proper** fraction is less than 1. An **improper** fraction is greater than or equal to 1.

5. Two fractions are **equivalent** if they have the same value.

7. Multiplying the numerator and denominator of a fraction by a number to obtain an equivalent fraction that involves larger numbers is called expressing the fraction in **higher** terms or **building** up the fraction.

Concepts

9. a. 2 is the common factor
   b. 3 is the common factor
   c. 5 is the common factor
   d. 7 is the common factor

11. This demonstrates equivalent fractions $\frac{2}{6} = \frac{1}{3}$.

13. a. The difference is the first example is factored while the second approach uses prime factorization.
    b. The results are the same.

15. This does not demonstrate factored numerator and denominator.

17. a. $8 = \frac{8}{1}$

    b. $-25 = -\frac{25}{1}$

    c. $x = \frac{x}{1}$

    d. $7a = \frac{7a}{1}$

Notation

19. $\frac{18}{24}=\frac{3\cdot3\cdot2}{3\cdot2\cdot2\cdot2}$

$=\frac{\cancel{3}\cdot3\cdot\cancel{2}}{\cancel{3}\cdot2\cdot2\cdot\cancel{2}}$

$=\frac{3}{4}$

Practice

21. $\frac{3}{9}=\frac{1}{3}$

23. $\frac{7}{21}=\frac{1}{3}$

25. $\frac{20}{30}=\frac{2}{3}$

27. $\frac{15}{6}=\frac{5}{2}$

29. $-\frac{28}{56}=-\frac{1}{2}$

31. $-\frac{90}{105}=-\frac{6}{7}$

33. $\frac{60}{108}=\frac{5}{9}$

35. $\frac{180}{210}=\frac{6}{7}$

37. $\frac{55}{67}$ is in lowest terms

39. $\frac{36}{96}=\frac{3}{8}$

41. $\frac{25x^2}{35x}=\frac{5x}{7}$

43. $\frac{12t}{15t}=\frac{4}{5}$

45. $\frac{6a}{7a}=\frac{6}{7}$

47. $\frac{7xy}{8xy}=\frac{7}{8}$

49. $-\frac{10rs}{30}=-\frac{rs}{3}$

51. $\frac{15st^3}{25xt^3}=\frac{3s}{5x}$

53. $\frac{35r^2t}{28rt^2}=\frac{5r}{4t}$

55. $\frac{56p^4}{28p^6}=\frac{2}{p^2}$

57. $\frac{7}{8}\left(\frac{5}{5}\right)=\frac{35}{40}$

59. $\frac{4}{5}\left(\frac{7}{7}\right)=\frac{28}{35}$

61. $\frac{5}{6}\left(\frac{9}{9}\right)=\frac{45}{54}$

63. $\frac{1}{2}\left(\frac{15}{15}\right)=\frac{15}{30}$

65. $\frac{2}{7}\left(\frac{2x}{2x}\right)=\frac{4x}{14x}$

67. $\frac{9}{10}\left(\frac{6t}{6t}\right)=\frac{54t}{60t}$

69. $\frac{5}{4s}\left(\frac{5}{5}\right)=\frac{25}{20s}$

71. $\frac{2}{15}\left(\frac{3y}{3y}\right)=\frac{6y}{45y}$

73. $3\left(\frac{5}{5}\right)=\frac{15}{5}$

75. $6\left(\frac{8}{8}\right) = \frac{48}{8}$

77. $4a\left(\frac{9}{9}\right) = \frac{36a}{9}$

79. $-2t\left(\frac{2}{2}\right) = -\frac{4t}{2}$

Applications

81. COMMUTING

The motorist has covered $\frac{3}{5}$ of the commute.

83. SINK HOLE

$-\frac{15}{16}$ inch

85. PERSONNEL RECORDS

Name	*Total Time*	*Time Alone*	*Amount Completed*
Bob	10 hours	7 hours	$\frac{7}{10}$
Ali	8 hours	1 hour	$\frac{1}{8}$

87. MUSIC

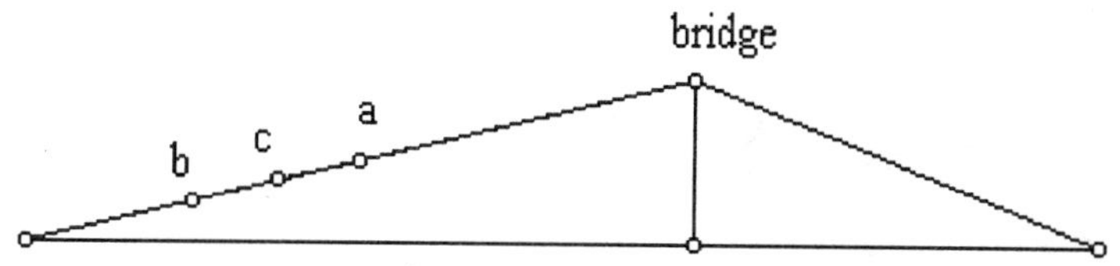

89. MACHINERY

There are two directions to turn the dial, one quarter turn to the left or three-quarter turn to the right.

91. SUPERMARKET DISPLAY

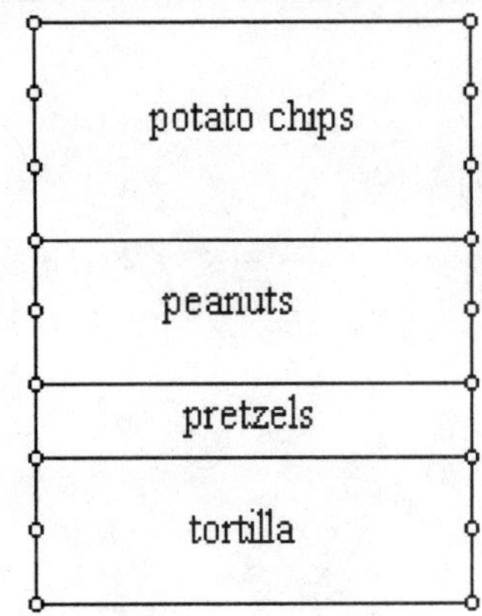

93. CAMERA

She should select a shutter speed of $\frac{1}{250}$.

Writing

95. Answers may vary
97. Answers may vary

Review

99. $$\begin{aligned} -5x+1&=16 \\ -5x&=15 \\ x&=-3 \end{aligned}$$

101. 564,112 to the nearest thousand is 564,000

103. The value of d dimes is $10d$.

Section 4.2 Multiplying Fractions

Vocabulary

1. The word *of* in mathematics usually means **multiply**.
3. The result of a multiplication problem is called the **product**.
5. In a triangle, b stands for the length of the **base** and h stands for the **height**.

Concepts

7. $\frac{a}{b} \cdot \frac{c}{d} = \frac{ac}{bd}$

9. a. $\frac{1}{4}$ will be shaded

   b. The rectangle is divided into 12 equal parts. One part has been shaded twice. $\frac{1}{3} \cdot \frac{1}{4} = \frac{1}{12}$

11. a. The product of numbers with unlike signs is negative.
    b. The product of two numbers with like signs is positive.

13. a. This is a true statement.
    b. This is a true statement.
    c. This is a false statement since $-\frac{3}{8}a = -\frac{3a}{8}$.
    d. This is a true statement.

Notation

15. $\frac{5}{8} \cdot \frac{7}{15} = \frac{5 \cdot 7}{8 \cdot 15}$

$$= \frac{5 \cdot 7}{8 \cdot 5 \cdot 3}$$

$$= \frac{\not{5} \cdot 7}{8 \cdot \not{5} \cdot 3}$$

$$= \frac{7}{24}$$

Practice

17. $\frac{1}{4} \bullet \frac{1}{2} = \frac{1}{8}$

19. $\frac{3}{8} \bullet \frac{7}{16} = \frac{21}{128}$

21. $\frac{2}{3} \bullet \frac{6}{7} = \frac{12}{21} = \frac{4}{7}$

23. $\frac{2(7)}{15} \bullet \frac{11}{2(4)} = \frac{77}{60}$

25. $-\frac{5(3)}{3(8)} \bullet \frac{8}{5(5)} = -\frac{1}{5}$

27. $\left(-\frac{11}{7(3)}\right)\left(-\frac{7(2)}{11(3)}\right) = \frac{2}{9}$

29. $\frac{7}{10}\left(\frac{10(2)}{7(3)}\right) = \frac{2}{3}$

31. $\frac{3}{4} \bullet \frac{4}{3} = 1$

33. $\frac{1}{3} \bullet \frac{3(5)}{4(4)} \bullet \frac{4}{5(5)} = \frac{1}{20}$

35. $\left(\frac{2}{3}\right)\left(-\frac{1}{2(2)(4)}\right)\left(-\frac{4}{5}\right) = \frac{1}{30}$

37. $\frac{5}{6} \bullet 18 = 5(3) = 15$

39. $15\left(-\frac{4}{5}\right) = 3(-4) = -12$

41. $\frac{5x}{12} \bullet \frac{1}{6x} = \frac{5}{72}$

43. $\frac{b}{12} \bullet \frac{3}{10b} = \frac{1}{40}$

45. $\frac{1}{3} \bullet 3d = d$

47. $\frac{2}{3} \bullet \frac{3s}{2} = s$

49. $-\frac{5}{6} \bullet \frac{6}{5} c = -c$

51. $\frac{x}{2} \bullet \frac{4}{9x} = \frac{2}{9}$

53. $4e \bullet \frac{e}{2} = 2e^2$

55. $\frac{5}{8x}\left(\frac{2x^3}{15}\right) = \frac{x^2}{12}$

57. $-\frac{5c}{6cd^2} \bullet \frac{12d^4}{c} = -\frac{10d^2}{c}$

59. $-\frac{4h^2}{5}\left(-\frac{15}{16h^3}\right) = \frac{3}{4h}$

61. $\frac{5}{6} \bullet x = \frac{5x}{6}$ or $\frac{5}{6}x$

63. $-\frac{8}{9} \bullet v = -\frac{8v}{9}$ or $-\frac{8}{9}v$

65. $\left(\frac{2}{3}\right)^2 = \frac{4}{9}$

67. $\left(-\frac{5}{9}\right)^2 = \frac{25}{81}$

69. $\left(\frac{4m}{3}\right)^2 = \frac{16m^2}{9}$

71. $\left(-\frac{3r}{4}\right)^3 = -\frac{27r^3}{64}$

73. Illustration 1

	$\frac{1}{2}$	$\frac{1}{3}$	$\frac{1}{4}$	$\frac{1}{5}$	$\frac{1}{6}$
$\frac{1}{2}$	$\frac{1}{4}$	$\frac{1}{6}$	$\frac{1}{8}$	$\frac{1}{10}$	$\frac{1}{12}$
$\frac{1}{3}$	$\frac{1}{6}$	$\frac{1}{9}$	$\frac{1}{12}$	$\frac{1}{15}$	$\frac{1}{18}$
$\frac{1}{4}$	$\frac{1}{8}$	$\frac{1}{12}$	$\frac{1}{16}$	$\frac{1}{20}$	$\frac{1}{24}$
$\frac{1}{5}$	$\frac{1}{10}$	$\frac{1}{15}$	$\frac{1}{20}$	$\frac{1}{25}$	$\frac{1}{30}$
$\frac{1}{6}$	$\frac{1}{12}$	$\frac{1}{18}$	$\frac{1}{24}$	$\frac{1}{30}$	$\frac{1}{36}$

75. Using $A=\frac{1}{2}bh, A=\frac{1}{2}(10)(3)=15\text{ ft}^2$

77. Using $A=\frac{1}{2}bh, A=\frac{1}{2}(3)(5)=\frac{15}{2}\text{ yd}^2$

Applications

79. THE CONSTITUTION

$\frac{2}{3}(435)=2(145)=290$

81. TENNIS BALL

$\frac{1}{3}(54)=18$ inches, bounce 1

$\frac{1}{3}(18)=6$ inches, bounce 2

$\frac{1}{3}(6)=2$ inches, bounce 3

83. COOKING

$\frac{1}{2}\left(\frac{3}{4}\right)=\frac{3}{8}$ cup of sugar, $\frac{1}{2}\left(\frac{1}{3}\right)=\frac{1}{6}$ cup of molasses

Note: the recipe is for 2 dozen so use ½ of each ingredient to make 1 dozen.

85. BOTANY

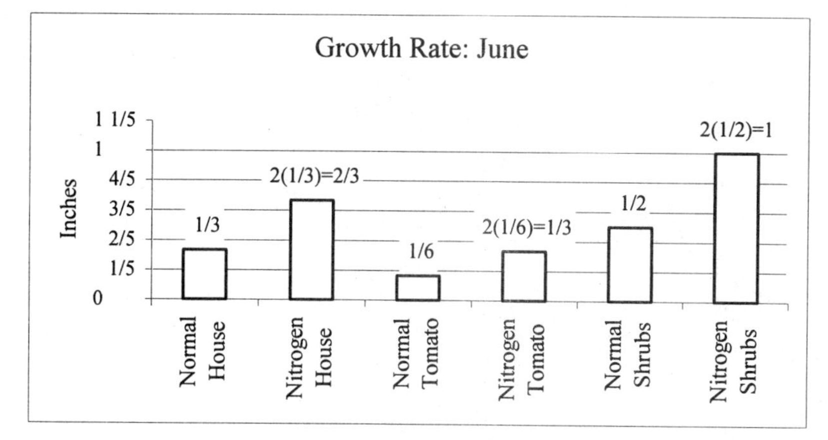

87. THE STARS AND STRIPES

Using $A=\frac{1}{2}bh$, $A=\frac{1}{2}(22)(11)=121 \text{ in}^2$

89. TILE DESIGN

Using $A=\frac{1}{2}bh$, $A=\frac{1}{2}(3)(3)=\frac{9}{2} \text{ in}^2$.

Now multiply this by four, $4\left(\frac{9}{4}\right)=18 \text{ in}^2$ of the tile is blue.

Writing

91. Answers may vary
93. Answers may vary

Review

95. $2(x+7)=2x+14$

97. $2(-6)+6\neq 6$, so $x=-6$ is not a solution

99. $125=5^3$

Section 4.3 Dividing Fractions

Vocabulary

1. Two numbers are called **reciprocals** if their product is 1.

Concepts

3. $\frac{1}{2} \div \frac{2}{3} = \frac{1}{2} \bullet \frac{3}{2}$

5. Dividing each rectangle into thirds represents $4 \div \frac{1}{3}$ and the quotient is 12.

7. The product of $\frac{4}{5}$ and it's reciprocal is one; $\frac{4}{5}\left(\frac{5}{4}\right) = 1$.

9. a. $15 \div 3 = 5$

   b. $15\left(\frac{1}{3}\right) = 5$

   c. Division by 3 is the same as multiplication by $\frac{1}{3}$.

Notation

11.

$$\begin{aligned} \frac{25}{36} \div \frac{10}{9} &= \frac{25}{36} \bullet \frac{9}{10} \\ &= \frac{25 \bullet 9}{36 \bullet 10} \\ &= \frac{5 \bullet 5 \bullet 9}{4 \bullet 9 \bullet 2 \bullet 5} \\ &= \frac{\cancel{5} \bullet 5 \bullet \cancel{9}}{4 \bullet \cancel{9} \bullet 2 \bullet \cancel{5}} \\ &= \frac{5}{8} \end{aligned}$$

Practice

13. $\frac{1}{2} \div \frac{3}{5} = \frac{1}{2} \bullet \frac{5}{3} = \frac{5}{6}$

15. $\frac{3}{16} \div \frac{1}{9} = \frac{3}{16} \bullet \frac{9}{1} = \frac{27}{16}$

17. $\frac{4}{5} \div \frac{4}{5} = \frac{4}{5} \bullet \frac{5}{4} = 1$

19. $\left(-\frac{7}{4}\right) \div \left(-\frac{21}{8}\right) = -\frac{7}{4} \bullet \left(-\frac{4(2)}{3(7)}\right) = \frac{2}{3}$

21. $3 \div \frac{1}{12} = 3(12) = 36$

23. $120 \div \frac{12}{5} = 120\left(\frac{5}{12}\right) = 50$

25. $-\frac{4}{5} \div (-6) = -\frac{4}{5} \bullet \left(\frac{1}{-6}\right) = \frac{2}{15}$

27. $\frac{15}{16} \div 180 = \frac{15}{16} \bullet \left(\frac{1}{12(15)}\right) = \frac{1}{192}$

29. $-\frac{9}{10} \div \frac{4}{15} = -\frac{9}{2(5)} \bullet \frac{3(5)}{4} = -\frac{27}{8}$

31. $\frac{9}{10} \div \left(-\frac{3}{25}\right) = \frac{3(3)}{2(5)} \bullet \left(-\frac{5(5)}{3}\right) = -\frac{15}{2}$

33. $-\frac{1}{8} \div 8 = -\frac{1}{8} \bullet \frac{1}{8} = -\frac{1}{64}$

35. $\frac{15}{32} \div \frac{15}{32} = \frac{15}{32} \bullet \frac{32}{15} = 1$

37. $\frac{4a}{5} \div \frac{3}{2} = \frac{4a}{5} \bullet \frac{2}{3} = \frac{8a}{15}$

39. $\frac{t}{8} \div \frac{3}{4} = \frac{t}{4(2)} \bullet \frac{4}{3} = \frac{t}{6}$

41. $\frac{13}{16b} \div \frac{1}{2} = \frac{13}{2(8)b} \bullet \frac{2}{1} = \frac{13}{8b}$

43. $-\frac{15}{32y} \div \frac{3}{4} = -\frac{3(5)}{(8)4y} \bullet \frac{4}{3} = -\frac{5}{8y}$

45. $a \div \frac{a}{b} = a \bullet \frac{b}{a} = b$

47. $\frac{x}{y} \div \frac{x}{y} = \frac{x}{y} \bullet \frac{y}{x} = 1$

49. $\frac{2s}{3t} \div (-6) = \frac{2s}{3t} \bullet \left(-\frac{1}{6}\right) = -\frac{s}{9t}$

51. $-\frac{9}{8}x \div \frac{3}{4x^2} = -\frac{3(3)}{2(4)}x \bullet \frac{4x^2}{3} = -\frac{3x^3}{2}$

53. $-8x \div \left(-\frac{4x^3}{9}\right) = -8x \div \left(-\frac{9}{4x^3}\right) = \frac{18}{x^2}$

55. $-\frac{x^2}{y^3} \div \frac{x}{y} = -\frac{x^2}{y^3} \bullet \frac{y}{x} = -\frac{x}{y^2}$

57. $-\frac{26x}{15} \div \frac{13}{45x} = -\frac{2(13x)}{15} \bullet \frac{3(15x)}{13} = -6x^2$

Applications

59. MARATHON

$26 \div \frac{1}{4} = 26(4) = 104$ laps

61. LASER TECHNOLOGY

$\frac{7}{8} \div \frac{1}{64} = \frac{7}{8} \bullet \frac{8^2}{1} = 56$ slices

63. UNDERGROUND CABLE

Route One: $15 \div \frac{3}{5} = 15\left(\frac{5}{3}\right) = 25$ days

Route Two: $12 \div \frac{2}{5} = 12\left(\frac{5}{2}\right) = 30$ days

The fewest installation days is Route one.

65. 3×5 CARDS

a. One inch on this ruler is divided into 16 parts.

b. The stack of cards is $\frac{3}{4}$ inch

c. One card is $\frac{3}{4} \div 90 = \frac{3}{4}\left(\frac{1}{90}\right) = \frac{1}{120}$ inch thick

67. FORESTRY

This map contains $6,284 \div \frac{4}{5} = 6,284\left(\frac{5}{4}\right) = 7,855$ sections .

Writing

69. Answers may vary
71. Answers may vary

Review

73. $4x + (-2) = -18$

$$4x = -16$$

$$x = -4$$

75. $p = r - c$ relates costs, profits, and revenue

77. This is a false statement.

79. $-3t + (-5T) + 4T + 8t = 5t - T$

Section 4.4 Adding and Subtracting Fractions

Vocabulary

1. The **least** common denominator for a set of fractions is the smallest number each denominator will divide exactly.
3. To express a fraction in **higher** terms, we multiply the numerator and denominator by the same number.

Concepts

5. This rule tells us how to add fractions having like **denominators**. To find the sum, we add the **numerators** and then write that result over the **common** denominator.

7. Since the denominators are not the same some preliminary work must be done.

9. a. The numerator and denominator are being multiplied by 4.
   b. The numerator and denominator are being multiplied by c.

11. a. a 5 appears once
    b. a 3 appears twice
    c. a 2 appears three times

13. The LCD for these two fractions is $2 \cdot 2 \cdot 3 \cdot 5 = 60$.

15. a. $\frac{1}{3} > \frac{1}{4}$
    b. $\frac{1}{3} = \frac{4}{12} > \frac{1}{4} = \frac{3}{12}$

Notation

17. 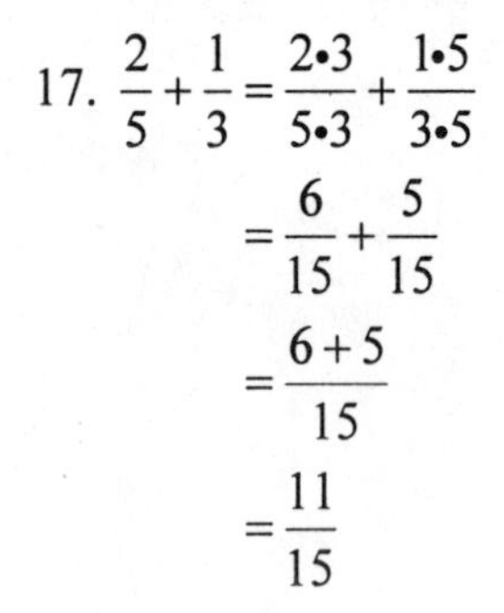

$$\begin{aligned}\frac{2}{5} + \frac{1}{3} &= \frac{2 \cdot 3}{5 \cdot 3} + \frac{1 \cdot 5}{3 \cdot 5} \\ &= \frac{6}{15} + \frac{5}{15} \\ &= \frac{6+5}{15} \\ &= \frac{11}{15}\end{aligned}$$

Practice

19. LCD (18, 6) = 2(3)(3) = 18
21. LCD (8, 6) = 2(2)(2)(3) = 24
23. LCD (8, 20) = 2(2)(2)(5) = 40
25. LCD (15, 12) = 2(2)(3)(5) = 60

27. $\frac{3}{7}+\frac{1}{7}=\frac{4}{7}$

29. $\frac{37}{103}-\frac{17}{103}=\frac{20}{103}$

31. $\frac{11}{25}-\frac{1}{25}=\frac{10}{25}=\frac{2}{5}$

33. $\frac{5}{d}+\frac{3}{d}=\frac{8}{d}$

35. $\frac{1}{4}+\frac{3}{8}=\frac{2}{8}+\frac{3}{8}=\frac{5}{8}$

37. $\frac{13}{20}-\frac{1}{5}=\frac{13}{20}-\frac{4}{20}=\frac{9}{20}$

39. $\frac{4}{5}+\frac{2}{3}=\frac{12}{15}+\frac{10}{15}=\frac{22}{15}$

41. $\frac{1}{8}+\frac{2}{7}=\frac{7}{56}+\frac{16}{56}=\frac{23}{56}$

43. $\frac{3}{4}-\frac{2}{3}=\frac{9}{12}-\frac{8}{12}=\frac{1}{12}$

45. $\frac{5}{6}-\frac{3}{4}=\frac{10}{12}-\frac{9}{12}=\frac{1}{12}$

47. $\frac{16}{25}-\left(-\frac{3}{10}\right)=\frac{32}{50}+\frac{15}{50}=\frac{47}{50}$

49. $-\frac{7}{16}+\frac{1}{4}=-\frac{7}{16}+\frac{4}{16}=-\frac{3}{16}$

51. $\frac{1}{12}-\frac{3}{4}=\frac{1}{12}-\frac{9}{12}=-\frac{8}{12}=-\frac{2}{3}$

53. $-\frac{5}{8}-\frac{1}{3}=-\frac{15}{24}-\frac{8}{24}=-\frac{23}{24}$

55. $-3+\frac{2}{5}=-\frac{15}{5}+\frac{2}{5}=-\frac{13}{5}$

57. $-\frac{3}{4}-5=-\frac{3}{4}-\frac{20}{4}=-\frac{23}{4}$

59. $\frac{7}{8}-\frac{t}{7}=\frac{49}{56}-\frac{8t}{56}=\frac{49-8t}{56}$

61. $\frac{4}{5}-\frac{2b}{9}=\frac{36}{45}-\frac{10b}{45}=\frac{36-10b}{45}$

63. $\frac{4}{7}-\frac{1}{r}=\frac{4r}{7r}-\frac{7}{7r}=\frac{4r-7}{7r}$

65. $-\frac{5}{9}+\frac{1}{y}=-\frac{5y}{9y}+\frac{9}{9y}=\frac{-5y+9}{9y}$

67. $\frac{1}{3}+\frac{1}{4}+\frac{1}{5}=\frac{20}{60}+\frac{15}{60}+\frac{12}{60}=\frac{47}{60}$

69. $-\frac{2}{3}+\frac{5}{4}+\frac{1}{6}=-\frac{8}{12}+\frac{15}{12}+\frac{2}{12}=\frac{9}{12}=\frac{3}{4}$

71. $\frac{5}{24}+\frac{3}{16}=\frac{10}{48}+\frac{9}{48}=\frac{19}{48}$

73. $-\frac{11}{15}-\frac{2}{9}=-\frac{33}{45}-\frac{10}{45}=-\frac{43}{45}$

75. $\frac{7}{25}+\frac{1}{15}=\frac{21}{75}+\frac{5}{75}=\frac{26}{75}$

77. $\frac{4}{27}+\frac{1}{6}=\frac{8}{54}+\frac{9}{54}=\frac{17}{54}$

79. $\frac{11}{60}-\frac{2}{45}=\frac{33}{180}-\frac{8}{180}=\frac{25}{180}=\frac{5}{36}$

81. $\frac{2}{15}-\frac{5}{12}=\frac{8}{60}-\frac{25}{60}=-\frac{17}{60}$

Applications

83. BOTANY

a. The growth was $\frac{5}{32}+\frac{1}{16}=\frac{5}{32}+\frac{2}{32}=\frac{7}{32}$ inches

b. The difference was $\frac{5}{32}-\frac{1}{16}=\frac{5}{32}-\frac{2}{32}=\frac{3}{32}$ inches

85. FAMILY DINNER

$\frac{3}{8}+\frac{2}{6}=\frac{9}{24}+\frac{8}{24}=\frac{17}{24}$ of the pizza was left.

They ate $\frac{5}{8}+\frac{4}{6}=\frac{15}{24}+\frac{16}{24}=\frac{31}{24}$ pizza, which is more than a whole pizza, so the family could not have been fed one pizza.

87. WEIGHTS AND MEASURES

The scale is off $\frac{1}{16}$ of a pound. This results in undercharging customers.

89. HIKING

From longest to shortest $\frac{4}{5},\frac{3}{4},\frac{5}{8}$, since $\frac{4}{5}=\frac{32}{40}$, $\frac{3}{4}=\frac{30}{40}$, and $\frac{5}{8}=\frac{25}{40}$.

91. STUDY HABITS

The fraction of students that study 2 hours or more are $\frac{2}{5}+\frac{3}{10}=\frac{4}{10}+\frac{3}{10}=\frac{7}{10}$.

93. GARAGE DOOR OPENER

The difference in strength is $\frac{1}{2}-\frac{1}{3}=\frac{3}{6}-\frac{2}{6}=\frac{1}{6}$ hp.

Writing

95. Answers may vary
97. Answers may vary

Review

99. $2(2+x)-3(x-1)$
$=4+2x-3x+3$
$=7-x$

101. $x-5$

103. $P=2l+2w$

LCM, GCF Study Set

1. LCM (3, 5) = 15
3. LCM (8, 14) = 56
5. LCM (14, 21) = 42
7. LCM (6, 18) = 18
9. LCM (44, 60) = 660
11. LCM (100, 120) = 660
13. LCM (6, 24, 36) = 72
15. LCM (18, 54, 63) = 378
17. GCF (6, 9) = 3
19. GCF (22, 33) = 11
21. GCF (16, 20) = 4
23. GCF (25, 100) = 25
25. GCF (100, 120) = 20
27. GCF (48, 108) = 12
29. GCF (18, 24, 36) = 6
31. GCF (18, 54, 63) = 9

33. NURSING
    2 hours is 120 minutes; LCM (45, 120) is 360 minutes or 6 hours

Section 4.5 Multiplying and Dividing Mixed Numbers

Vocabulary

1. A **mixed** number is the sum of a whole number and a proper fraction.
3. To **graph** a number means to locate its position on a number line and highlight it using a heavy dot.

Concepts

5. a. $-5\frac{1}{2}^{0}$

   b. $-1\frac{7}{8}$ in

7. a. The arrow is registering $-2\frac{2}{3}$.

   b. It will register $-3\frac{1}{3}$.

9. $-\frac{4}{5}, -\frac{2}{5}, \frac{1}{5}$

11. DIVING

    Forward $2\frac{1}{2}$ somersaults from the pike position.

13.

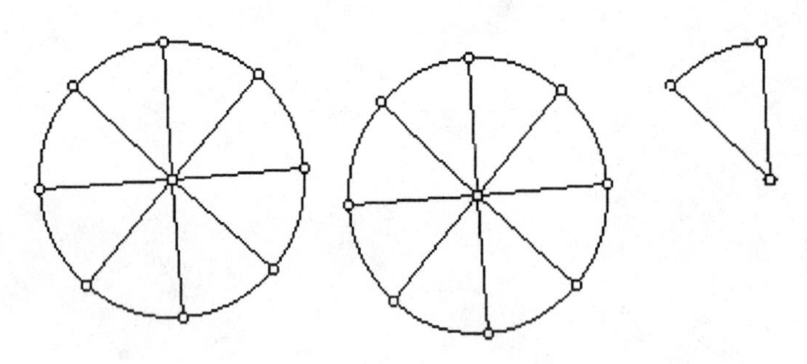

Notation

15. $-5\frac{1}{4}\bullet 1\frac{1}{7} = -\frac{21}{4}\bullet\frac{8}{7}$

$$= -\frac{21\bullet 8}{4\bullet 7}$$

$$= -\frac{\cancel{7}\bullet 3\bullet \cancel{4}\bullet 2}{\cancel{4}\bullet \cancel{7}}$$

$$= -\frac{6}{1}$$

$$= -6$$

Practice

17. $\frac{15}{4} = 3\frac{3}{4}$

19. $\frac{29}{5} = 5\frac{4}{5}$

21. $-\frac{20}{6} = -3\frac{2}{6} = -3\frac{1}{3}$

23. $\frac{127}{12} = 10\frac{7}{12}$

25. $6\frac{1}{2} = \frac{13}{2}$

27. $20\frac{4}{5} = \frac{104}{5}$

29. $-6\frac{2}{9} = -\frac{56}{9}$

31. $200\frac{2}{3} = \frac{602}{3}$

33. $\left\{-2\frac{8}{9}, 1\frac{2}{3}, \frac{16}{5}\right\}$

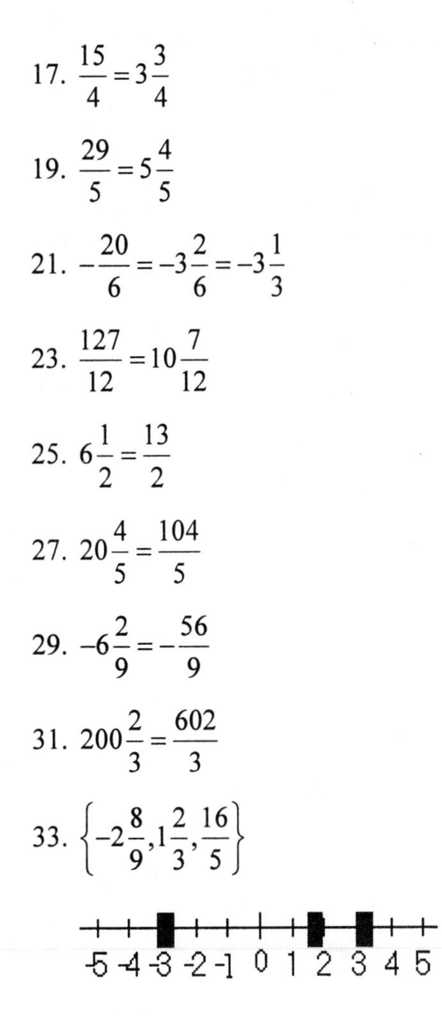

35. $\left\{-\frac{10}{3}, -\frac{98}{99}, 3\frac{1}{7}\right\}$

-5 -4 -3 -2 -1 0 1 2 3 4 5

37. $1\frac{2}{3}\cdot 2\frac{1}{7} = \frac{5}{3}\cdot\frac{15}{7} = \frac{5}{3}\cdot\frac{3\cdot 5}{7} = \frac{25}{7} = 3\frac{4}{7}$

39. $-7\frac{1}{2}\left(-1\frac{2}{5}\right) = \frac{15}{2}\cdot\frac{7}{5} = \frac{3\cdot 5}{2}\cdot\frac{7}{5} = \frac{21}{2} = 10\frac{1}{2}$

41. $3\frac{1}{16}\cdot 4\frac{4}{7} = \frac{49}{16}\cdot\frac{32}{7} = \frac{7\cdot 7}{16}\cdot\frac{16\cdot 2}{7} = 14$

43. $-6\cdot 2\frac{7}{24} = -\frac{6}{1}\cdot\frac{55}{24} = -\frac{6}{1}\cdot\frac{55}{6\cdot 4} = -\frac{55}{4} = -13\frac{3}{4}$

45. $2\frac{1}{2}\left(-3\frac{1}{3}\right) = -\frac{5}{2}\cdot\frac{10}{3} = -\frac{5}{2}\cdot\frac{5\cdot 2}{3} = -\frac{25}{3} = -8\frac{1}{3}$

47. $2\frac{5}{8}\cdot\frac{5}{27} = \frac{21}{8}\cdot\frac{5}{27} = \frac{3\cdot 7}{8}\cdot\frac{5}{3\cdot 9} = \frac{35}{72}$

49. $\left(1\frac{2}{3}\right)(6)\left(-\frac{1}{8}\right) = \left(\frac{5}{3}\right)\left(\frac{6}{1}\right)\left(-\frac{1}{8}\right) = -\frac{30}{24} = -\frac{5}{4} = -1\frac{1}{4}$

51. $\left(1\frac{2}{3}\right)^2 = \left(\frac{5}{3}\right)^2 = \frac{25}{9} = 2\frac{7}{9}$

53. $\left(-1\frac{1}{3}\right)^3 = \left(-\frac{4}{3}\right)^3 = -\frac{64}{27} = -2\frac{10}{27}$

55. $3\frac{1}{3}\div 1\frac{5}{6} = \frac{10}{3}\div\frac{11}{6} = \frac{10}{3}\cdot\frac{6}{11} = \frac{20}{11} = 1\frac{9}{11}$

57. $-6\frac{3}{5}\div 7\frac{1}{3} = -\frac{33}{5}\div\frac{22}{3} = -\frac{33}{5}\cdot\frac{3}{22} = -\frac{9}{10}$

59. $-20\frac{1}{4} \div \left(-1\frac{11}{16}\right)$

$= -\frac{81}{4} \div \left(-\frac{27}{16}\right)$

$= \frac{81}{4} \bullet \frac{16}{27}$

$= \frac{3^4}{4} \bullet \frac{4^2}{3^3}$

$= 12$

61. $6\frac{1}{4} \div 20 = \frac{25}{4} \bullet \frac{1}{20}$

$= \frac{5^2}{4} \bullet \frac{1}{4(5)}$

$= \frac{5}{16}$

63. $1\frac{2}{3} \div \left(-2\frac{1}{2}\right) = \frac{5}{3} \bullet -\frac{2}{5}$

$= -\frac{2}{3}$

65. $8 \div \left(3\frac{1}{5}\right) = \frac{8}{1} \bullet \frac{5}{16}$

$= \frac{8}{1} \bullet \frac{5}{2(8)}$

$= \frac{5}{2}$

$= 2\frac{1}{2}$

67. $-4\frac{1}{2} \div 2\frac{1}{4} = -\frac{9}{2} \bullet \frac{4}{9} = -2$

Applications

69. CALORIES

$$20\left(3\frac{1}{5}\right)=\frac{20}{1}\bullet\frac{16}{5}=64\text{ calories}$$

71. SHOPPING

$$4\frac{1}{4}\bullet\frac{64}{100}=\frac{17}{4}\bullet\frac{4^3}{4(25)}=\frac{68}{25}=\$2.72\quad\left(\text{Note: }\$0.64=\frac{64}{100}\right)$$

73. SUBDIVISION

The remaining acreage is 1,000 – 100 = 900 acres.

$$900\div 1\frac{1}{3}=\frac{900}{1}\bullet\frac{3}{4}=\frac{225(4)}{1}\bullet\frac{3}{4}=675\text{ lots}$$

75. GRAPH PAPER

The dimensions of this paper are $11\left(\frac{1}{4}\right)=\frac{11}{4}=2\frac{3}{4}$ inches by $5\left(\frac{1}{4}\right)=\frac{5}{4}=1\frac{1}{4}$ inches.

77. EMERGENCY EXIT

$$A=\frac{1}{2}bh$$

$$A=\frac{1}{2}\left(8\frac{1}{4}\right)\left(10\frac{1}{3}\right)$$

$$A=\frac{1}{2}\left(\frac{3(11)}{4}\right)\left(\frac{31}{3}\right)$$

$$A=\frac{341}{8}$$

$$A=42\frac{5}{8}\text{ in}^2$$

79. FIRE ESCAPE

The total height is $43(105)=4{,}515$ inches, so the total number of steps is

$$4{,}515\div 7\frac{1}{2}=4{,}515\bullet\frac{2}{15}=301(2)=602\text{ steps}.$$

81. SHOPPING ON THE INTERNET

She should choose size 14, slim cut.

Writing

83. Answers may vary
85. Answers may vary

Review

87. $3^2 \cdot 2^3 = 72$
89. 4(8)
91. Division by 2 must be undone.
93. The variables are different.

Section 4.6 Adding and Subtracting Mixed Numbers

Vocabulary

1. By the **commutative** property of addition, we can add numbers in any order.

3. To do the subtraction, we **borrow** 1 in the form of $\frac{3}{3}$.

Concepts

5. a. The whole number part is 76 the fractional part is $\frac{3}{4}$.

   b. $76+\frac{3}{4}$

7. The fundamental property of fractions is being highlighted.

9. a. $9\frac{17}{16}=9+\frac{16}{16}+\frac{1}{16}=10\frac{1}{16}$

   b. $1{,}288\frac{7}{3}=1{,}288+\frac{6}{3}+\frac{1}{3}=1{,}290\frac{1}{3}$

   c. $16\frac{12}{8}=16+\frac{8}{8}+\frac{4}{8}=17\frac{1}{2}$

   d. $45\frac{24}{20}=45+\frac{20}{20}+\frac{4}{20}=46\frac{1}{5}$

Notation

11.

$$\begin{aligned}70\frac{3}{5}+39\frac{2}{7}&=70+\frac{3}{5}+39+\frac{2}{7}\\&=70+39+\frac{3}{5}+\frac{2}{7}\\&=109+\frac{3}{5}+\frac{2}{7}\\&=109+\frac{3\cdot 7}{5\cdot 7}+\frac{2\cdot 5}{7\cdot 5}\\&=109+\frac{21}{35}+\frac{10}{35}\\&=109+\frac{31}{35}\\&=109\frac{31}{35}\end{aligned}$$

Practice

13. $2\frac{1}{5}+2\frac{1}{5}=4\frac{2}{5}$

15. $8\frac{2}{7}-3\frac{1}{7}=5\frac{1}{7}$

17. $3\frac{1}{4}+4\frac{1}{4}=7\frac{2}{4}=7\frac{1}{2}$

19. $4\frac{1}{6}+1\frac{1}{5}=5+\frac{5}{30}+\frac{6}{30}=5\frac{11}{30}$

21. $2\frac{1}{2}-1\frac{1}{4}=2\frac{2}{4}-1\frac{1}{4}=1\frac{1}{4}$

23. $2\frac{5}{6}-1\frac{3}{8}=2\frac{20}{24}-1\frac{9}{24}=1\frac{11}{24}$

25. $5\frac{1}{2}+3\frac{4}{5}=5\frac{5}{10}+3\frac{8}{10}=8\frac{13}{10}=8+\frac{10}{10}+\frac{3}{10}=9\frac{3}{10}$

27. $7\frac{1}{2}-4\frac{1}{7}=7\frac{7}{14}-4\frac{2}{14}=3\frac{5}{14}$

29. $56\frac{2}{5}+73\frac{1}{3}=129+\frac{6}{15}+\frac{5}{15}=129\frac{11}{15}$

31. $380\frac{1}{6}+17\frac{1}{4}=397+\frac{2}{12}+\frac{3}{12}=397\frac{5}{12}$

33. $228\frac{5}{9}+44\frac{2}{3}=272+\frac{15}{27}+\frac{18}{27}=272\frac{33}{27}=273\frac{6}{27}=273\frac{2}{9}$

35. $778\frac{5}{7}-155\frac{1}{3}=778\frac{15}{21}-155\frac{7}{21}=623\frac{8}{21}$

37. $140\frac{5}{6}-129\frac{4}{5}=140\frac{25}{30}-129\frac{24}{30}=11\frac{1}{30}$

39. $422\frac{13}{16}-321\frac{3}{8}=422\frac{13}{16}-321\frac{6}{16}=101\frac{7}{16}$

41. $16\frac{1}{4}-13\frac{3}{4}=15\frac{5}{4}-13\frac{3}{4}=2\frac{2}{4}=2\frac{1}{2}$

43. $76\frac{1}{6}-49\frac{7}{8}=76\frac{4}{24}-49\frac{21}{24}=75\frac{28}{24}-49\frac{21}{24}=26\frac{7}{24}$

45. $140\frac{3}{16}-129\frac{3}{4}=140\frac{3}{16}-129\frac{12}{16}=139\frac{19}{16}-129\frac{12}{16}=10\frac{7}{16}$

47. $334\frac{1}{9}-13\frac{5}{6}=334\frac{2}{18}-13\frac{15}{18}=333\frac{20}{18}-13\frac{15}{18}=320\frac{5}{18}$

49. $7-\frac{2}{3}=6\frac{3}{3}-\frac{2}{3}=6\frac{1}{3}$

51. $9-8\frac{3}{4}=8\frac{4}{4}-8\frac{3}{4}=\frac{1}{4}$

53. $4\frac{1}{7}-\frac{4}{5}=4\frac{5}{35}-\frac{28}{35}=3\frac{40}{35}-\frac{28}{35}=3\frac{12}{35}$

55. $6\frac{5}{8}-3=3\frac{5}{8}$

57. $\frac{7}{3}+2=2\frac{1}{3}+2=4\frac{1}{3}$

59. $2+1\frac{7}{8}=3\frac{7}{8}$

61. $12\frac{1}{2}+5\frac{3}{4}+35\frac{1}{6}=52+\frac{6}{12}+\frac{9}{12}+\frac{2}{12}=52+\frac{17}{12}=53\frac{5}{12}$

63. $58\frac{7}{8}+340+61\frac{1}{4}=459+\frac{7}{8}+\frac{2}{8}=459+\frac{9}{8}=460\frac{1}{8}$

65. $-3\frac{3}{4}+\left(-1\frac{1}{2}\right)=-3\frac{3}{4}+\left(-1\frac{2}{4}\right)=-4\frac{5}{4}=-5\frac{1}{4}$

67. $-4\frac{5}{8}-1\frac{1}{4}=-4\frac{5}{8}-1\frac{2}{8}=-5\frac{7}{8}$

Applications

69. FREEWAY TRAVEL

Grand Avenue and Citrus Avenue are $3\frac{1}{2}-\frac{3}{4}=3\frac{2}{4}-\frac{3}{4}=2\frac{6}{4}-\frac{3}{4}=2\frac{3}{4}$ miles apart.

71. TRAIL MIX

The recipe will yield $2\frac{3}{4}+2\left(\frac{1}{2}\right)+\frac{2}{3}+\frac{1}{3}+2\frac{2}{3}+\frac{1}{4}=5+\frac{9}{12}+\frac{8}{12}+\frac{4}{12}+\frac{8}{12}+\frac{3}{12}$

$$=5+\frac{32}{12}$$

$$=7\frac{8}{12}$$

$$=7\frac{2}{3}\text{ cups}$$

73. HOSE REPAIR

The hose is $50-1\frac{1}{2}=49\frac{2}{2}-1\frac{1}{2}=48\frac{1}{2}$ feet long.

75. SHIPPING

a.

	Rate	*Time*	*Distance = Rate (Time)*
Passenger	$16\frac{1}{2}$	1	$16\frac{1}{2}$
Cargo	$5\frac{1}{5}$	1	$5\frac{1}{5}$

b. At 1:00 a.m. the ships are $16\frac{1}{2}+5\frac{1}{5}=21+\frac{5}{10}+\frac{2}{10}=21\frac{7}{10}$ miles apart.

77. SERVICE STATION

a. The difference in price is $179\frac{9}{10}-159\frac{9}{10}=20$ cents.

b. The cost per gallon for the full serve is 30 cents more, $199\frac{9}{10}-169\frac{9}{10}=30$.

79. WATER SLIDE

The slide was $311\frac{5}{12}-119\frac{3}{4}=311\frac{5}{12}-119\frac{9}{12}$

$=310\frac{17}{12}-119\frac{9}{12}$

$=191\frac{8}{12}$

$=191\frac{2}{3}$ feet long.

Writing

81. Answers may vary
83. Answers may vary

Review

85. $2x-1=3x-8$

$7=x$

87. $8(x+2)+3(2-x)$

$=8x+16+6-3x$

$=5x+22$

89. $-2-(-8)=-2+8=6$

91. Area measures the amount of surface a figure encloses.

Section 4.7 Order of Operations and Complex Fractions

Vocabulary

1. $\dfrac{\frac{1}{2}}{\frac{3}{4}}$ is a **complex** fraction.

Concepts

3. $\dfrac{\frac{2}{3}}{\frac{1}{5}} = \dfrac{2}{3} \div \dfrac{1}{5}$

5. The common denominator of all the fractions is 15.
7. When this complex fraction is simplified the result will be negative.
9. $\text{LCD}(4,5,6) = 60$

Notation

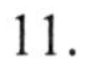

11.

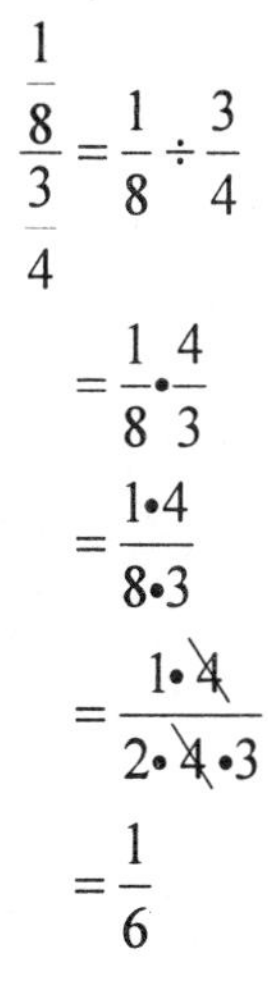

$$\frac{\frac{1}{8}}{\frac{3}{4}} = \frac{1}{8} \div \frac{3}{4}$$

$$= \frac{1}{8} \bullet \frac{4}{3}$$

$$= \frac{1 \bullet 4}{8 \bullet 3}$$

$$= \frac{1 \bullet \cancel{4}}{2 \bullet \cancel{4} \bullet 3}$$

$$= \frac{1}{6}$$

Practice

13. $\frac{2}{3}\left(-\frac{1}{4}\right)+\frac{1}{2}=-\frac{1}{6}+\frac{3}{6}=\frac{2}{6}=\frac{1}{3}$

15. $\frac{4}{5}-\left(-\frac{1}{3}\right)^2=\frac{4}{5}-\frac{1}{9}$

$$=\frac{36}{45}-\frac{5}{45}$$

$$=\frac{31}{45}$$

17. $-4\left(-\frac{1}{5}\right)-\left(\frac{1}{4}\right)\left(-\frac{1}{2}\right)=\frac{4}{5}+\frac{1}{8}$

$$=\frac{32}{40}+\frac{5}{40}$$

$$=\frac{37}{40}$$

19. $1\frac{3}{5}\left(\frac{1}{2}\right)^2\left(\frac{3}{4}\right)=\frac{8}{5}\left(\frac{1}{4}\right)\left(\frac{3}{4}\right)$

$$=\frac{3}{10}$$

21.

$$\frac{7}{8}-\left(\frac{4}{5}+1\frac{3}{4}\right)=\frac{7}{8}-\left(\frac{4}{5}+\frac{7}{4}\right)$$

$$=\frac{7}{8}-\left(\frac{16}{20}+\frac{35}{20}\right)$$

$$=\frac{7}{8}-\frac{51}{20}$$

$$=\frac{35}{40}-\frac{102}{40}$$

$$=-\frac{67}{40}$$

$$=-1\frac{27}{40}$$

23.

$$\left(\frac{9}{20}\div 2\frac{2}{5}\right)+\left(\frac{3}{4}\right)^2=\left(\frac{9}{20}\div\frac{12}{5}\right)+\left(\frac{9}{16}\right)$$

$$=\left(\frac{9}{20}\bullet\frac{5}{12}\right)+\left(\frac{9}{16}\right)$$

$$=\frac{3}{16}+\frac{9}{16}$$

$$=\frac{12}{16}$$

$$=\frac{3}{4}$$

25. $$\left(-\frac{3}{4}\bullet\frac{9}{16}\right)+\left(\frac{1}{2}-\frac{1}{8}\right)=-\frac{27}{64}+\frac{3}{8}$$

$$=-\frac{27}{64}+\frac{24}{64}$$

$$=-\frac{3}{64}$$

27.

$$\left|\frac{2}{3}-\frac{9}{10}\right|\div\left(-\frac{1}{5}\right)=\left|\frac{20}{30}-\frac{27}{30}\right|\bullet(-5)$$

$$=\left|-\frac{7}{30}\right|\bullet(-5)$$

$$=\frac{7}{30}\bullet(-5)$$

$$=-\frac{7}{6}$$

$$=-1\frac{1}{6}$$

29.

$$\left(2-\frac{1}{2}\right)^2+\left(2+\frac{1}{2}\right)^2=\left(\frac{3}{2}\right)^2+\left(\frac{5}{2}\right)^2$$
$$=\frac{9}{4}+\frac{25}{4}$$
$$=\frac{34}{4}$$
$$=\frac{17}{2}$$
$$=8\frac{1}{2}$$

31. $\frac{1}{2}(-7)=-\frac{7}{2}$

$$\left(-\frac{7}{2}\right)^2=\frac{49}{4}$$

33. $\frac{1}{2}\left(\frac{11}{2}\right)=\frac{11}{4}$

$$\left(\frac{11}{4}\right)^2=\frac{121}{16}$$

35.

$$a=1\frac{3}{4};b=-\frac{1}{5};r=-1\frac{2}{3};c=-\frac{2}{3}$$
$$\frac{1}{3}b^2+c=\frac{1}{3}\left(-\frac{1}{5}\right)^2+\left(-\frac{2}{3}\right)$$
$$\frac{1}{3}b^2+c=\frac{1}{3}\left(\frac{1}{25}\right)+\left(-\frac{2}{3}\right)$$
$$\frac{1}{3}b^2+c=\left(\frac{1}{75}\right)+\left(-\frac{2}{3}\right)$$
$$\frac{1}{3}b^2+c=\left(\frac{1}{75}\right)+\left(-\frac{50}{75}\right)$$
$$\frac{1}{2}b^2+c=-\frac{49}{75}$$

37.

$$a = 1\frac{3}{4}; b = -\frac{1}{5}; r = -1\frac{2}{3}; c = -\frac{2}{3}$$

$$-1 - ar = -1 - 1\frac{3}{4}\left(-1\frac{2}{3}\right)$$

$$-1 - ar = -1 - \frac{7}{4}\left(-\frac{5}{3}\right)$$

$$-1 - ar = -1 + \frac{35}{12}$$

$$-1 - ar = \frac{23}{12}$$

$$-1 - ar = 1\frac{11}{12}$$

39.

$$P = 2l + 2w$$

$$P = 2\left(2\frac{7}{8}\right) + 2\left(1\frac{1}{4}\right)$$

$$P = 2\left(\frac{23}{8}\right) + 2\left(\frac{5}{4}\right)$$

$$P = \frac{23}{4} + \frac{5}{2}$$

$$P = \frac{23}{4} + \frac{10}{4}$$

$$P = \frac{33}{4}$$

$$P = 8\frac{1}{4} \text{ inches}$$

41. $$\frac{\frac{2}{3}}{\frac{4}{5}} = \frac{2}{3} \bullet \frac{5}{4} = \frac{5}{6}$$

43. $$\frac{-\frac{14}{15}}{\frac{7}{10}} = -\frac{14}{15} \bullet \frac{10}{7} = -\frac{4}{3} = -1\frac{1}{3}$$

45. $\dfrac{5}{\frac{10}{21}} = 5 \bullet \dfrac{21}{10} = \dfrac{21}{2} = 10\dfrac{1}{2}$

47. $\dfrac{-\frac{5}{6}}{-1\frac{7}{8}} = \dfrac{5}{6} \bullet \left(\dfrac{8}{15}\right) = \dfrac{5}{2 \bullet 3} \bullet \left(\dfrac{4 \bullet 2}{3 \bullet 5}\right) = \dfrac{4}{9}$

49. $\dfrac{\frac{1}{2}+\frac{1}{4}}{\frac{1}{2}-\frac{1}{4}} = \dfrac{\frac{3}{4}}{\frac{1}{4}} = \dfrac{3}{4} \bullet \dfrac{4}{1} = 3$

51. $\dfrac{\frac{3}{8}+\frac{1}{4}}{\frac{3}{8}-\frac{1}{4}} = \dfrac{\frac{3}{8}+\frac{2}{8}}{\frac{3}{8}-\frac{2}{8}} = \dfrac{\frac{5}{8}}{\frac{1}{8}} = \dfrac{5}{8} \bullet \dfrac{8}{1} = 5$

53. $\dfrac{\frac{1}{5}+3}{-\frac{4}{25}} = \dfrac{3\frac{1}{5}}{-\frac{4}{25}} = \dfrac{\frac{16}{5}}{-\frac{4}{25}} = \dfrac{16}{5} \bullet \left(-\dfrac{25}{4}\right) = -20$

55. $\dfrac{5\frac{1}{2}}{-\frac{1}{4}+\frac{3}{4}} = \dfrac{\frac{11}{2}}{\frac{1}{2}} = \dfrac{11}{2} \bullet \dfrac{2}{1} = 11$

57. $\dfrac{\frac{1}{5}-\left(-\frac{1}{4}\right)}{\frac{1}{4}+\frac{4}{5}} = \dfrac{\frac{1}{5}+\frac{1}{4}}{\frac{5}{20}+\frac{16}{20}} = \dfrac{\frac{4}{20}+\frac{5}{20}}{\frac{21}{20}} = \dfrac{\frac{9}{20}}{\frac{21}{20}} = \dfrac{9}{20} \bullet \dfrac{20}{21} = \dfrac{3}{7}$

59. $\dfrac{\frac{1}{3}+\left(-\frac{5}{6}\right)}{1\frac{1}{3}} = \dfrac{\frac{2}{6}+\left(-\frac{5}{6}\right)}{\frac{4}{3}} = \dfrac{-\frac{1}{2}}{\frac{4}{3}} = -\dfrac{1}{2} \bullet \dfrac{3}{4} = -\dfrac{3}{8}$

61. $x=-\frac{3}{4}; y=\frac{7}{8}$

$$\frac{x+y}{2}=\frac{-\frac{3}{4}+\frac{7}{8}}{2}=\left(-\frac{6}{8}+\frac{7}{8}\right)\left(\frac{1}{2}\right)=\frac{1}{8}\left(\frac{1}{2}\right)=\frac{1}{16}$$

63. $x=-\frac{3}{4}; y=\frac{7}{8}$

$$\left|\frac{2x}{y-x}\right|=\left|\frac{2\left(-\frac{3}{4}\right)}{\frac{7}{8}-\left(-\frac{3}{4}\right)}\right|=\left|\frac{\left(-\frac{3}{2}\right)}{\frac{7}{8}-\left(-\frac{6}{8}\right)}\right|=\left|\frac{\left(-\frac{3}{2}\right)}{\frac{13}{8}}\right|=\left|-\frac{3}{2}\bullet\frac{8}{13}\right|=\left|-\frac{12}{13}\right|=\frac{12}{13}$$

Applications

65. SANDWICH SHOP

$$\frac{1\frac{3}{4}+2\frac{1}{2}}{\frac{1}{2}}=\frac{\frac{7}{4}+\frac{5}{2}}{\frac{1}{2}}=\frac{\frac{7}{4}+\frac{10}{4}}{\frac{1}{2}}=\frac{\frac{17}{4}}{\frac{1}{2}}=\frac{17}{4}\bullet 2=\frac{17}{2}=8\frac{1}{2} \text{ sandwiches}$$

67. PHYSICAL FITNESS

	Rate	*Time*	*Distance = Rate(Time)*
Jogger	$2\frac{1}{2}$	$1\frac{1}{2}$	$2\frac{1}{2}\left(1\frac{1}{2}\right)=\frac{5}{2}\left(\frac{3}{2}\right)=\frac{15}{4}=3\frac{3}{4}$
Cyclist	$7\frac{1}{5}$	$1\frac{1}{2}$	$7\frac{1}{5}\left(1\frac{1}{2}\right)=\frac{36}{5}\left(\frac{3}{2}\right)=\frac{54}{5}=10\frac{4}{5}$

The total distance apart is $3\frac{3}{4}+10\frac{4}{5}=13+\frac{3}{4}+\frac{4}{5}=13+\frac{15}{20}+\frac{16}{20}=13+\frac{31}{20}=14\frac{11}{20}$ miles .

69. POSTAGE RATES

$\frac{1}{16}+\frac{5}{8}+3\left(\frac{1}{16}\right)=\frac{1}{16}+\frac{10}{16}+\frac{3}{16}=\frac{14}{16}=\frac{7}{8}$ oz which is less than one ounce.

It can be mailed for the 1 oz rate.

71. PHYSICAL THERAPY

The total distance was

$$7\left(\frac{1}{4}\right)+7\left(\frac{1}{2}\right)+7\left(\frac{3}{4}\right)=\frac{7}{4}+\frac{7}{2}+\frac{21}{4}=\frac{7}{4}+\frac{14}{4}+\frac{21}{4}=\frac{42}{4}=\frac{21}{2}=10\frac{1}{2}\text{ miles}.$$

73. AMUSEMENT PARK

$$\frac{1}{\frac{1}{10}+\frac{1}{15}}=\frac{1}{\frac{3}{30}+\frac{2}{30}}=\frac{1}{\frac{5}{30}}=\frac{30}{5}=6\text{ seconds}$$

Writing

75. Answers may vary
77. Answers may vary

Review

79. $-4d-(-7d)=-4d+7d=3d$

81. $2x(-x)$

83. $2+3[-3-(-4-1)]=2+3[-3-(-5)]$
$=2+3[-3+5]$
$=2+3[2]$
$=8$

85. $3\bullet3\bullet3\bullet x\bullet x\bullet x\bullet x\bullet x=27x^5$

Section 4.8 Solving Equations Containing Fractions

Vocabulary

1. To find the **reciprocal** of a fraction, invert the numerator and the denominator.
3. The **least common denominator** of a set of fractions is the smallest number each denominator will divide exactly.

Concepts

5. Let $x = 40$, then $\frac{5}{8}x = \frac{5}{8}(40) = 5^2 = 25$, so this is a solution.
7. The product of reciprocals is one.
9. a. $\frac{4}{5}p$

   b. $\frac{1}{4}t$

11. Both sides of the equation can be multiplied by $\frac{3}{2}$ or multiply both sides by three and then divide by two.

Notation

13.
$$\frac{7}{8}x = 21$$
$$\frac{8}{7}\left(\frac{7}{8}x\right) = \frac{8}{7}(21)$$
$$x = 24$$

15. a. This is a **true** statement.
    b. The is a **false** statement.
    c. This is a **true** statement.
    d. This is a **true** statement.

Practice

17. $\frac{4}{7}x = 16$

$$x = \frac{7}{4}(16)$$

$$x = 28$$

19. $\frac{7}{8}t = -28$

$$x = \frac{8}{7}(-28)$$

$$x = -32$$

21. $-\frac{3}{5}h = 4$

$$h = -\frac{5}{3}(4)$$

$$h = -\frac{20}{3}$$

23. $\frac{2}{3}x = \frac{4}{5}$

$$x = \frac{3}{2}\left(\frac{4}{5}\right)$$

$$x = \frac{6}{5}$$

25. $\frac{2}{5}y = 0$

$$y = \frac{5}{2}(0)$$

$$y = 0$$

27. $-\frac{5c}{6} = -25$

$$c = -\frac{6}{5}(-25)$$

$$c = 30$$

29. $\dfrac{-5f}{7} = -2$

$$c = -\frac{7}{5}(-2)$$

$$c = \frac{14}{5}$$

31. $\dfrac{5}{8}y = \dfrac{1}{10}$

$$y = \frac{8}{5}\left(\frac{1}{10}\right)$$

$$y = \frac{4}{25}$$

33. $2x + 1 = 0$

$$2x = -1$$

$$x = -\frac{1}{2}$$

35. $5x - 1 = 1$

$$5x = 2$$

$$x = \frac{2}{5}$$

37. $6x = 2x - 11$

$$4x = -11$$

$$x = -\frac{11}{4}$$

39. $2(y - 3) = 7$

$$2y - 6 = 7$$

$$2y = 13$$

$$y = \frac{13}{2}$$

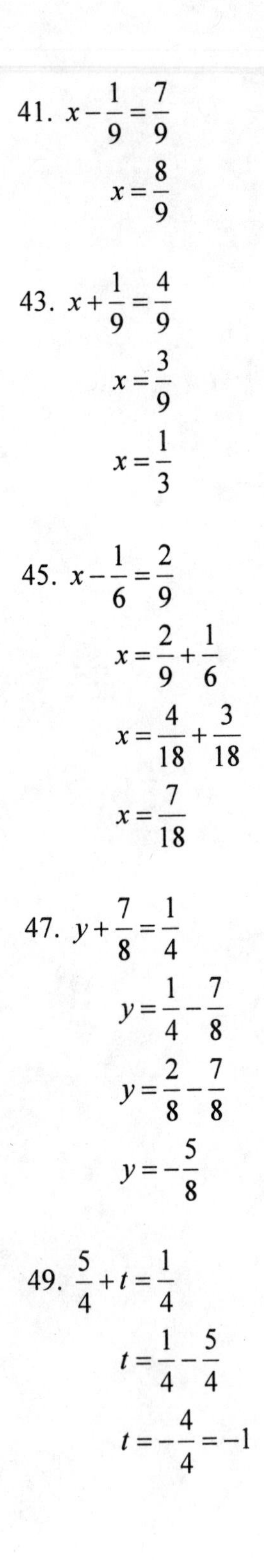

41. $x-\frac{1}{9}=\frac{7}{9}$

$x=\frac{8}{9}$

43. $x+\frac{1}{9}=\frac{4}{9}$

$x=\frac{3}{9}$

$x=\frac{1}{3}$

45. $x-\frac{1}{6}=\frac{2}{9}$

$x=\frac{2}{9}+\frac{1}{6}$

$x=\frac{4}{18}+\frac{3}{18}$

$x=\frac{7}{18}$

47. $y+\frac{7}{8}=\frac{1}{4}$

$y=\frac{1}{4}-\frac{7}{8}$

$y=\frac{2}{8}-\frac{7}{8}$

$y=-\frac{5}{8}$

49. $\frac{5}{4}+t=\frac{1}{4}$

$t=\frac{1}{4}-\frac{5}{4}$

$t=-\frac{4}{4}=-1$

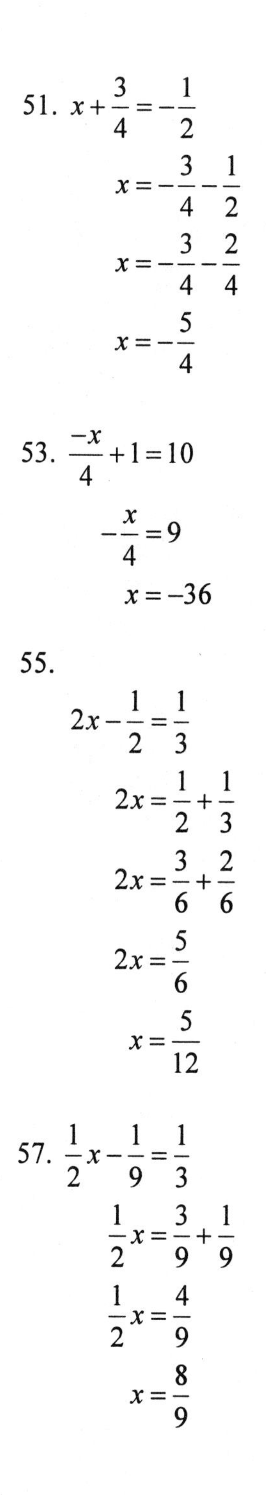

51. $x+\frac{3}{4}=-\frac{1}{2}$

$$x=-\frac{3}{4}-\frac{1}{2}$$

$$x=-\frac{3}{4}-\frac{2}{4}$$

$$x=-\frac{5}{4}$$

53. $\frac{-x}{4}+1=10$

$$-\frac{x}{4}=9$$

$$x=-36$$

55.

$$2x-\frac{1}{2}=\frac{1}{3}$$

$$2x=\frac{1}{2}+\frac{1}{3}$$

$$2x=\frac{3}{6}+\frac{2}{6}$$

$$2x=\frac{5}{6}$$

$$x=\frac{5}{12}$$

57. $\frac{1}{2}x-\frac{1}{9}=\frac{1}{3}$

$$\frac{1}{2}x=\frac{3}{9}+\frac{1}{9}$$

$$\frac{1}{2}x=\frac{4}{9}$$

$$x=\frac{8}{9}$$

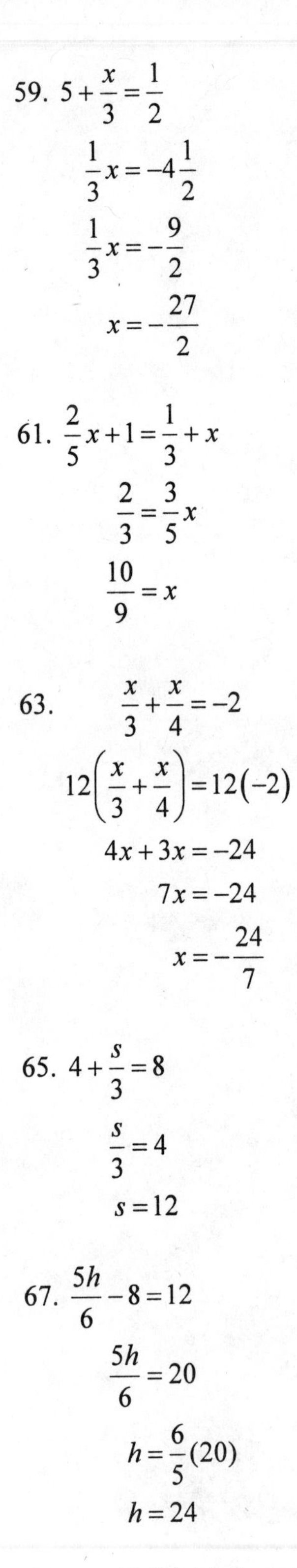

59. $5+\frac{x}{3}=\frac{1}{2}$

$$\frac{1}{3}x=-4\frac{1}{2}$$

$$\frac{1}{3}x=-\frac{9}{2}$$

$$x=-\frac{27}{2}$$

61. $\frac{2}{5}x+1=\frac{1}{3}+x$

$$\frac{2}{3}=\frac{3}{5}x$$

$$\frac{10}{9}=x$$

63. $\frac{x}{3}+\frac{x}{4}=-2$

$$12\left(\frac{x}{3}+\frac{x}{4}\right)=12(-2)$$

$$4x+3x=-24$$

$$7x=-24$$

$$x=-\frac{24}{7}$$

65. $4+\frac{s}{3}=8$

$$\frac{s}{3}=4$$

$$s=12$$

67. $\frac{5h}{6}-8=12$

$$\frac{5h}{6}=20$$

$$h=\frac{6}{5}(20)$$

$$h=24$$

69. $-4+9+\frac{5t}{12}=0$

$$\frac{5t}{12}=-5$$

$$t=\frac{12}{5}(-5)$$

$$t=-12$$

71. $-3-2+\frac{4x}{15}=0$

$$\frac{4x}{15}=5$$

$$y=\frac{15}{4}(5)$$

$$y=\frac{75}{4}$$

Applications

73. TRANSMISSION REPAIR

Analyze the problem

- Only $\frac{1}{3}$ of the customers needed new transmissions.
- The shop installed 32 new transmissions last year.
- Find the number of customers the shop had last year.

Form an equation

Let x = the number of customers last year

Key phrase: one-third of ***Translation:*** multiply

$\frac{1}{3}$ of the number of customers last year was 32.

$$\frac{1}{3}x=32$$

Solve the equation

$$\frac{1}{3}x=32$$

$$3\left(\frac{1}{3}x\right)=3(32)$$

$$x=96$$

State the conclusion The shop had 96 customers last year.

Check the result If we find 1/3 of 96, we get 32. The answer checks.

75. TOOTH DEVELOPMENT

Let t represent teeth

$$\frac{4}{5}t = 16$$

$$\frac{5}{4}\left(\frac{4}{5}t\right) = \frac{5}{4}(16)$$

$$t = 20$$

The child will eventually have 20 teeth.

77. HOME SALES

Let h represent homes, 9 homes represent the remaining one-fourth of the homes.

$$\frac{1}{4}h = 9$$

$$h = 4(9)$$

$$h = 36$$

There are 36 homes in the subdivision.

79. TELEPHONE BOOK

Let p represent total pages, 150 is one-third of the phone book.

$$\frac{1}{3}p = 150$$

$$p = 450$$

There are 450 pages in the telephone book.

81. SAFETY REQUIREMENT

Let w represent width

$$A = lw$$

$$30 = 3\frac{3}{4}w$$

$$30 = \frac{15}{4}w$$

$$\frac{4}{15}(30) = w$$

$$8 = w$$

The taillight must be 8 inches wide.

83. CPR CLASS

Let c represent the time for the complete class

$$\frac{1}{4}c+\frac{2}{3}c+30=c$$

$$\frac{3}{12}c+\frac{8}{12}c+30=c$$

$$30=\frac{1}{12}c$$

$$360=c$$

The course is 360 minutes.

Writing

85. Answers may vary
87. Answers may vary

Review

89. $a(b+c)=ab+ac$

91. 41 degrees Fahrenheit is 5 degrees Celsius.

$$C=\frac{5(F-32)}{9}=\frac{5(9)}{9}=5$$

93. $5x-3=2x+12$

$$3x=15$$

$$x=5$$

95. 12,590,767 to the nearest million is 13,000,000.

Chapter 4 Key Concept

1. $\frac{15}{25} = \frac{15 \div 5}{25 \div 5} = \frac{3}{5}$

3. $\frac{1}{5} = \frac{1 \cdot 7}{5 \cdot 7} = \frac{7}{35}$

Chapter 4 Review

Section 4.1 The Fundamental Property of Fractions

1. This woman spends $\frac{7}{24}$ of her day sleeping.

3. $\frac{2}{-3}=-\frac{2}{3}=\frac{-2}{3}$

5. The numerator and denominator are divided by two.

7. a. $\frac{15}{45}=\frac{1}{3}$

   b. $\frac{20}{48}=\frac{5}{12}$

   c. $-\frac{63x^2}{84x}=-\frac{63x}{84}=-\frac{3x}{4}$

   d. $\frac{66m^3n}{108m^4n}=\frac{66}{108m}=\frac{11}{18m}$

9. a. $\frac{2}{3}\bullet\frac{6}{6}=\frac{12}{18}$

   b. $-\frac{3}{8}\bullet\frac{2}{2}=-\frac{6}{16}$

   c. $\frac{7}{15}\bullet\frac{3a}{3a}=\frac{21a}{45a}$

   d. $\frac{4}{1}\bullet\frac{9}{9}=\frac{36}{9}$

Section 4.2 Multiplying Fractions

11. a. $\frac{3}{4}x=\frac{3x}{4}$ is a true statement.

    b. $-\frac{5}{9}e\neq-\frac{5}{9e}$, so it is a false statement.

13. a. $\left(\frac{3}{4}\right)^2 = \frac{9}{16}$

b. $\left(-\frac{5}{2}\right)^3 = -\frac{125}{8}$

c. $\left(\frac{x}{3}\right)^2 = \frac{x^2}{9}$

d. $\left(-\frac{2c}{5}\right)^3 = -\frac{8c^3}{125}$

15. $A = \frac{1}{2}bh$

$A = \frac{1}{2}(15)(8)$

$A = 60 \text{ in}^2$

Section 4.3 Dividing Fractions

17. a. $\frac{1}{6} \div \frac{11}{25} = \frac{1}{6} \bullet \frac{25}{11} = \frac{25}{66}$

b. $-\frac{7}{8} \div \frac{1}{4} = -\frac{7}{8} \bullet \frac{4}{1} = -\frac{7}{2}$

c. $-\frac{15}{16} \div (-10) = -\frac{15}{16} \bullet \left(-\frac{1}{10}\right) = \frac{3}{32}$

d. $8 \div \frac{16}{5} = 8 \bullet \frac{5}{16} = \frac{5}{2}$

e. $\frac{t}{8} \div \frac{1}{4} = \frac{t}{8} \bullet \frac{4}{1} = \frac{t}{2}$

f. $\frac{4a}{5} \div \frac{a}{2} = \frac{4a}{5} \bullet \frac{2}{a} = \frac{8}{5}$

g. $-\frac{a}{b} \div -\frac{b}{a} = -\frac{a}{b} \bullet \left(-\frac{a}{b}\right) = \frac{a^2}{b^2}$

h. $\frac{2}{3}x \div \left(-\frac{x^2}{9}\right) = \frac{2}{3}x \bullet \left(-\frac{9}{x^2}\right) = -\frac{6}{x}$

Section 4.4 Adding and Subtracting Fractions

19. a. $\frac{2}{7}+\frac{3}{7}=\frac{5}{7}$

b. $-\frac{3}{5}-\frac{3}{5}=-\frac{6}{5}$

c. $\frac{3}{x}-\frac{1}{x}=\frac{2}{x}$

d. $\frac{7}{8}+\frac{t}{8}=\frac{7+t}{8}$

21. $30=2{\cdot}3{\cdot}5$

$45=3{\cdot}3{\cdot}5$

$\text{LCD}(30,45)=2{\cdot}3{\cdot}3{\cdot}5=90$

23. MACHINE SHOP

$\frac{3}{4}-\frac{17}{32}=\frac{24}{32}-\frac{17}{32}=\frac{7}{32}$ inch must be milled away.

Section 4.5 Multiplying and Dividing Fractions

25. a. $2\frac{1}{6}$

b. $\frac{13}{6}$

27. a. $9\frac{3}{8}=\frac{75}{8}$

b. $-2\frac{1}{5}=-\frac{11}{5}$

c. $100\frac{1}{2}=\frac{201}{2}$

d. $1\frac{99}{100}=\frac{199}{100}$

29. a. $-5\frac{1}{4}\bullet\frac{2}{35}=-\frac{21}{4}\bullet\frac{2}{35}=-\frac{3}{10}$

b. $\left(-3\frac{1}{2}\right)\div\left(-3\frac{2}{3}\right)=-\frac{7}{2}\div\left(-\frac{11}{3}\right)=-\frac{7}{2}\bullet\left(-\frac{3}{11}\right)=\frac{21}{22}$

c. $\left(-6\frac{2}{3}\right)(-6)=\frac{20}{3}\left(\frac{6}{1}\right)=40$

d. $-8\div 3\frac{1}{5}=-8\div\left(\frac{16}{5}\right)=-8\bullet\left(\frac{5}{16}\right)=-\frac{5}{2}=-2\frac{1}{2}$

Section 4.6 Adding and Subtracting Mixed Numbers

31. a. $1\frac{3}{8}+2\frac{1}{5}=3+\frac{3}{8}+\frac{1}{5}=3+\frac{15}{40}+\frac{8}{40}=3\frac{23}{40}$

b. $3\frac{1}{2}+2\frac{2}{3}=5+\frac{1}{2}+\frac{2}{3}=5+\frac{3}{6}+\frac{4}{6}=5\frac{7}{6}=6\frac{1}{6}$

c. $2\frac{5}{6}-1\frac{3}{4}=2\frac{10}{12}-1\frac{9}{12}=1\frac{1}{12}$

d. $3\frac{7}{16}-2\frac{1}{8}=3\frac{7}{16}-2\frac{2}{16}=1\frac{5}{16}$

33. a. $133\frac{1}{9}+49\frac{1}{6}=182+\frac{1}{9}+\frac{1}{6}=182+\frac{2}{18}+\frac{3}{18}=182\frac{5}{18}$

b. $98\frac{11}{20}+14\frac{3}{5}=112+\frac{11}{20}+\frac{3}{5}=112+\frac{11}{20}+\frac{12}{20}=112\frac{23}{20}=113\frac{3}{20}$

c. $50\frac{5}{8}-19\frac{1}{6}=50\frac{15}{24}-19\frac{4}{24}=31\frac{11}{24}$

d. $375\frac{3}{4}-59=316\frac{3}{4}$

Section 4.7 Order of Operations and Complex Fractions

35. a. $\frac{3}{4}+\left(-\frac{1}{3}\right)^2\left(\frac{5}{4}\right)=\frac{3}{4}+\left(\frac{1}{9}\right)\left(\frac{5}{4}\right)=\frac{3}{4}+\frac{5}{36}=\frac{27}{36}+\frac{5}{36}=\frac{32}{36}=\frac{8}{9}$

b. $\left(\frac{2}{3}\div\frac{16}{9}\right)-\left(1\frac{2}{3}\bullet\frac{1}{15}\right)=\left(\frac{2}{3}\bullet\frac{9}{16}\right)-\left(\frac{5}{3}\bullet\frac{1}{15}\right)$

$=\left(\frac{3}{8}\right)-\left(\frac{1}{9}\right)$

$=\frac{27}{72}-\frac{8}{72}$

$=\frac{19}{72}$

37. a.

$c=-\frac{3}{4}, d=\frac{1}{8}, e=-2\frac{1}{16}$

$d^2-2c=\left(\frac{1}{8}\right)^2-2\left(-\frac{3}{4}\right)$

$=\frac{1}{64}+\frac{3}{2}$

$=\frac{1}{64}+\frac{96}{64}$

$=\frac{97}{64}$

b.

$c=-\frac{3}{4}, d=\frac{1}{8}, e=-2\frac{1}{16}$

$-cd+e=-\left(-\frac{3}{4}\right)\left(\frac{1}{8}\right)-2\frac{1}{16}$

$=\frac{3}{32}-\frac{33}{16}$

$=\frac{3}{32}-\frac{66}{32}$

$=-\frac{63}{32}$

c.

$$c=-\frac{3}{4}, d=\frac{1}{8}, e=-2\frac{1}{16}$$

$$e\div(cd)=-2\frac{1}{16}\div\left(-\frac{3}{4}\right)\left(\frac{1}{8}\right)$$

$$=\frac{33}{16}\div\frac{3}{32}$$

$$=\frac{33}{16}\bullet\frac{32}{3}$$

$$=22$$

d.

$$c=-\frac{3}{4}, d=\frac{1}{8}, e=-2\frac{1}{16}$$

$$\frac{c-d}{e}=\frac{-\frac{3}{4}-\frac{1}{8}}{-2\frac{1}{16}}$$

$$=\frac{-\frac{6}{8}-\frac{1}{8}}{-\frac{33}{16}}$$

$$=\frac{-\frac{7}{8}}{-\frac{33}{16}}$$

$$=-\frac{7}{8}\bullet\left(-\frac{16}{33}\right)$$

$$=\frac{14}{33}$$

Section 4.8 Solving Equations Containing Fractions

39. a.

$$\frac{c}{3}-\frac{c}{8}=2$$

$$\frac{8c}{24}-\frac{3c}{24}=2$$

$$\frac{5c}{24}=2$$

$$c=\left(\frac{24}{5}\right)(2)$$

$$c=\frac{48}{5}$$

b. $\frac{5h}{9}-1=-3$

$$\frac{5h}{9}=-2$$

$$h=\frac{9}{5}(-2)$$

$$h=-\frac{18}{5}$$

c. $4-\frac{d}{4}=0$

$$4=\frac{d}{4}$$

$$16=d$$

d.

$$\frac{t}{10}-\frac{2}{3}=\frac{1}{5}$$

$$\frac{t}{10}=\frac{2}{3}+\frac{1}{5}$$

$$t=10\left(\frac{10}{15}+\frac{3}{15}\right)$$

$$t=10\left(\frac{13}{15}\right)$$

$$t=\frac{26}{3}$$

Chapter 4 Test

1. a. $\frac{4}{5}$ of the plant is above ground.

   b. $\frac{1}{5}$ of the plant is below ground.

3. $-\frac{3x}{4}\left(\frac{1}{5x^2}\right)=-\frac{3}{4}\left(\frac{1}{5x}\right)=-\frac{3}{20x}$

5. $\frac{4a}{3}\div\frac{a^2}{9}=\frac{4a}{3}\bullet\frac{9}{a^2}=\frac{12}{a}$

7. $\frac{7}{8}\bullet\frac{3a}{3a}=\frac{21a}{24a}$

9. SPORTS CONTRACT

   $13\frac{1}{2}\div 9=\frac{27}{2}\bullet\frac{1}{9}=\frac{3}{2}=\$1\frac{1}{2}$ million/year

11. $157\frac{5}{9}+103\frac{3}{4}=260+\frac{5}{9}+\frac{3}{4}$

$$=260+\frac{20}{36}+\frac{27}{36}$$

$$=260+1+\frac{11}{36}$$

$$=261\frac{11}{36}$$

13. BOXING

    a. The difference in weight was 0 pounds.

    b. The chest difference was $42\frac{1}{4}-39\frac{1}{2}=41\frac{5}{4}-39\frac{2}{4}=2\frac{3}{4}$ inches.

    c. The waist difference was $31\frac{3}{4}-28=3\frac{3}{4}$ inches.

15. SEWING

$$10\frac{1}{2}+2\left(\frac{5}{8}\right)=10\frac{1}{2}+\frac{5}{4}$$
$$=10\frac{2}{4}+\frac{5}{4}$$
$$=11\frac{3}{4}\text{ inches}$$

17.

$$\left(\frac{2}{3}\bullet\frac{5}{16}\right)-\left(-1\frac{3}{5}\div4\frac{4}{5}\right)=\left(\frac{5}{24}\right)-\left(-\frac{8}{5}\div\frac{24}{5}\right)$$
$$=\left(\frac{5}{24}\right)-\left(-\frac{8}{5}\bullet\frac{5}{24}\right)$$
$$=\left(\frac{5}{24}\right)-\left(-\frac{1}{3}\right)$$
$$=\frac{5}{24}+\frac{8}{24}$$
$$=\frac{13}{24}$$

19. $$\frac{\frac{1}{2}+\frac{1}{3}}{-\frac{1}{6}-\frac{1}{3}}=\frac{\frac{3}{6}+\frac{2}{6}}{-\frac{1}{6}-\frac{2}{6}}=\frac{\frac{5}{6}}{-\frac{3}{6}}=\frac{5}{6}\left(-\frac{6}{3}\right)=-\frac{5}{3}$$

21. $$6x-4=-3$$
$$6x=1$$
$$x=\frac{1}{6}$$

23. JOB APPLICANTS

One fourth of the applicants, p, did not have experience, or 36 people.

$$\frac{1}{4}p=36$$
$$p=144\text{ people applied}$$

25. The product refers to multiplication, so when multiplying reciprocals the result is one.

Chapters 1 – 4 Cumulative Review Exercises

1. 5, 434, 700

3. THE STOCK MARKET
   The highest mark was approximately 11, 555 at 10:30 a.m.

5. $4{,}679 + 3{,}457 = 8{,}136$

7. $5{,}345 \times 56 = 299{,}320$

9. The pool perimeter is $2(150) + 2(75) = 250$ feet .

11. $84 = 2^2 \bullet 3 \bullet 7$

13. $360 = 2^3 \bullet 3^2 \bullet 5$

15. $6 + (-2)(-5) = 6 + 10 = 16$

17. $\dfrac{2(-7)+3(2)}{2(-2)} = \dfrac{-14+6}{-4} = \dfrac{-8}{-4} = 2$

19. $x + 15$

21. $4x$

23. $x = 4$
    $2x - 1 = 2(4) - 1 = 7$

25. $x = 4$
    $3x - x^3 = 3(4) - 4^3 = 12 - 64 = -52$

27. $-3(5x) = -15x$

29. $-2(3x - 4) = -6x + 8$

31. $-3x + 8x = 5x$

33. $4x - 3y - 5x + 2y = -x - y$

35. $3x + 2 = -13$
    $3x = -15$
    $x = -5$

37. $\dfrac{y}{4} - 1 = -5$
    $\dfrac{y}{4} = -4$
    $y = -16$

39. $6x - 12 = 2x + 4$
$$4x = 16$$
$$x = 4$$

41. OBSERVATION HOURS

The student must observe $100 - 37 = 63$ hours, or $\frac{63}{3} = 21$ three-hour sessions.

43. $\frac{21}{28} = \frac{3}{4}$

45. $\frac{6}{5}\left(-\frac{2}{3}\right) = -\frac{4}{5}$

47. $\frac{2}{3} + \frac{3}{4} = \frac{8}{12} + \frac{9}{12} = \frac{17}{12} = 1\frac{5}{12}$

49. $3\frac{5}{6} = \frac{23}{6}$

51. $4\frac{2}{3} + 5\frac{1}{4} = 9 + \frac{2}{3} + \frac{1}{4}$
$$= 9 + \frac{8}{12} + \frac{3}{12}$$
$$= 9\frac{11}{12}$$

53. FIRE HAZARD

The distance between the ground terminal and the hot terminal increased by $\frac{3}{4} - \frac{1}{16} = \frac{12}{16} - \frac{1}{16} = \frac{11}{16}$ inch.

55. $\left(\frac{1}{4} - \frac{7}{8}\right) \div \left(-2\frac{3}{16}\right) = \left(\frac{2}{8} - \frac{7}{8}\right) \div \left(-\frac{35}{16}\right)$
$$= \left(-\frac{5}{8}\right) \cdot \left(-\frac{16}{35}\right)$$
$$= \frac{2}{7}$$

57. $x+\frac{1}{5}=-\frac{14}{15}$

$x=-\frac{17}{15}$

Check

$-\frac{17}{15}+\frac{3}{15}=-\frac{14}{15}$

59. $\frac{2}{3}x=-10$

$x=-15$

Check

$\frac{2}{3}(-15)=-10$

61. The difference is the equal sign in an equation.

Section 5.1 An Introduction to Decimals

Vocabulary

1. From left to right, **thousands, hundreds, tens, ones, decimal point, tenths, hundredths, thousandths, and ten thousandths.**
3. We can approximate a decimal number using the process called **rounding**.

Concepts

5. a. 32.415 is thirty-two and four hundred fifteen thousandths.
   b. The whole number part is 32.
   c. The fractional part is $\frac{415}{1{,}000}$.
   d. In expanded notation $30+2+\frac{4}{10}+\frac{1}{100}+\frac{5}{1{,}000}$.

7. From left to right $\left\{-3\frac{1}{100}, -0.7, \frac{7}{10}, 3.01\right\}$

-10 -9 -8 -7 -6 -5 -4 -3 -2 -1 0 1 2 3 4 5 6 7 8 9 10

9. a. True
   b. False
   c. True
   d. True

11. The shaded part is $\frac{47}{100}$ or 0.47.

13.

0.3

Notation

15. 9,816.0245

Practice

17. fifty and one tenth; $50\frac{1}{10}$

19. negative one hundred thirty-seven ten thousandths; $-\frac{137}{10,000}$

21. three hundred four and three ten thousandths; $304\frac{3}{10,000}$

23. negative seventy-two and four hundred ninety-three thousandths; $-72\frac{493}{1,000}$

25. −0.39
27. 6.187
29. 506.1
31. 2.7
33. −0.14
35. 33.00
37. 3.142
39. 1.414
41. 39
43. 2,988
45. a. Nearest dollar $3,090
    b. Nearest ten cents $3,090.30

47. $-23.45 < -23.1$
49. $-0.065 > -0.066$
51. From least to greatest $132.64, 132.6401, 132.6499$.

Applications

53. WRITING A CHECK
    The amount is $1,025.78.

55. INJECTIONS
    The arrow should be placed two increments to the left of 0.4.

57. METRIC SYSTEM
    a. 0.30
    b. 1,609.34
    c. 453.59
    d. 3.79

59. GEOLOGY

Sample	Location	Size	Classification
A	Riverbank	0.009	Sand
B	Pond	0.0007	Silt
C	NE corner	0.095	Granule
D	Dry lake	0.00003	Clay

61. AIR QUALITY
From the highest reading to the smallest: Texas City; Houston; Westport; Galveston; White Plains; Crestline.

63. OLYMPICS
The gold went to Retton, the silver to Szabo, and the bronze to Pauca.

65. E-COMMERCE
For the third quarter of 1997 the loss was approximately - $0.07 and for the last quarter of 1998 the loss was approximately - $0.30.

Writing

67. Answers may vary.
69. Answers may vary.
71. Answers may vary.

Review

73. $75\frac{3}{4}+88\frac{4}{5}=75\frac{15}{20}+88\frac{16}{20}=163\frac{31}{20}=164\frac{11}{20}$

75. $5R-3(6-R)=5R-18+3R=8R-18$

77. $A=\frac{1}{2}(16)(9)$

$A=72\text{ in}^2$

79. $-2+(-3)+4=-5+4=-1$

Section 5.2 Addition and Subtraction with Decimals

Vocabulary

1. The answer to an addition problem is called the **sum**.
3. Every whole number has an unwritten decimal **point** to its right.

Concepts

5. a. $0.3+0.17=0.47$

   b. $0.3=\frac{3}{10};0.17=\frac{17}{100}$

   $\frac{3}{10}+\frac{17}{100}=\frac{30}{100}+\frac{17}{100}=\frac{47}{100}$

   c. $\frac{47}{100}=0.47$

   d. The results are the same.

Practice

7. $32.5+7.4=39.9$

9. $21.6+33.12=54.72$

11. $12+3.9=15.9$

13. $0.03034+0.2003=0.23064$

15. $247.9+40+0.56=288.46$

17. $45+9.9+0.12+3.02=58.04$

19. $12.98-3.45=9.53$

21. $78.1-7.81=70.29$

23. $5-0.023=4.977$

25. $24-23.81=0.19$

27. $-45.6+34.7=-10.9$

29. $46.09+(-7.8)=38.29$

31. $-7.8+(-6.5)=-14.3$

33. $-0.0045+(-0.031)=-0.0355$

35. $-9.5-7.1=-16.6$

37. $30.03-(-17.88)=47.91$

39. $-2.002-(-4.6)=2.598$

41. $-7-(-18.01)=11.01$

43. $3.4-6.6+7.3=4.1$

45. $(-9.1-6.05)-(-51)=35.85$

47. $16-(67.2+6.27)=-57.47$

49. $(-7.2+6.3)-(-3.1-4)=6.2$

51. $|-14.1+6.9|+8=|-7.2|+8=15.2$

53. $2.43+5.6=8.03$

Applications

55. SPORTS PAGE
   a. The Italian team's time was $53.03+0.014=53.044$ seconds.
   b. The second place score was $102.71-0.33=102.38$ points.

57. VEHICLE SPECIFICATIONS
The wheelbase of this car is $187.8-(43.5+40.9)=187.8-84.4=103.4$ inches.

59. BAROMETRIC PRESSURE
The difference between the areas of high and low pressure is 30.7 – 28.9 = 1.8.
We could expect fair weather in Texas.

61. OFFSHORE DRILLING

	Underwater	Underground	Total
Design 1	1.74 miles	2.32 miles	1.74 + 2.32 = 4.06 miles
Design 2	2.90 miles	0 miles	2.90 + 0 = 2.90 miles

63. AMERICAN RECORDHOLDERS
The difference in the two times is $54.48-10.49=43.99$ seconds.

65. DEPOSIT SLIP
The subtotal is $242.50+116.10+47.93+359.16=\$765.69$.
The total deposit is $\$765.69-25=\$740.69$.

67. THE HOME SHOPPING NETWORK
   a. The difference of the MSRP and the sale price is $149.79-47.85=\$101.94$.
   b. The set will cost $47.85+7.95=\$55.80$.

69. $2,367.909+5,789.0253=8,156.9343$

71. $9,000.09-7,067.445=1,932.645$

73. $3,434.768-(908-2.3+0.0098)=3,434.768-905.7098=2,529.0582$

Writing

75. Answers may vary.
77. Answers may vary.

Review

79. $44\frac{3}{8}+66\frac{1}{5}=44\frac{15}{40}+66\frac{8}{40}=110\frac{23}{40}$

81. $-\frac{15}{26}\bullet 1\frac{4}{9}=-\frac{15}{26}\bullet\frac{13}{9}=-\frac{5}{6}$

Section 5.3 Multiplication with Decimals

Vocabulary

1. In the multiplication problem $2.89 \cdot 15.7$, the numbers 2.89 and 15.7 are called **factors**. The answer 45.373, is called the **product**.

Concepts

3. To multiply decimals, multiply them as if they were **whole** numbers. The number of decimal places in the product is the same as the **sum** of the decimal places of the factors.
5. When we move the decimal point to the right the decimal gets **larger**.

7. a. $\frac{3}{10} \cdot \frac{7}{100} = \frac{21}{1,000}$

   b. $\frac{3}{10} = 0.3; \frac{7}{100} = 0.07$

   $0.3(0.07) = 0.021 = \frac{21}{1,000}$

Practice

9. $0.4(0.2) = 0.08$

11. $-0.5(0.3) = -0.15$

13. $1.4(0.7) = 0.98$

15. $0.08(0.9) = 0.072$

17. $-5.6(-2.2) = 12.32$

19. $-4.9(0.001) = -0.0049$

21. $-0.35(0.24) = -0.084$

23. $-2.13(4.05) = -8.6265$

25. $16(0.6) = 9.6$

27. $-7(8.1) = -56.7$

29. $0.04(306) = 12.24$

31. $60.61(-0.3) = -18.183$

33. $-0.2(0.3)(-0.4) = 0.024$

35. $5.5(10)(-0.3) = -16.5$

37. $4.2(10) = 42$

39. $67.164(100) = 6,716.4$

41. $-0.056(10) = -0.56$

43. $1,000(8.05) = 8,050$

45. $0.098(10,000) = 980$

47. $-0.2(1,000) = -200$

49.

Decimal	Its Square
0.1	0.01
0.2	0.04
0.3	0.09
0.4	0.16
0.5	0.25
0.6	0.36
0.7	0.49
0.8	0.64
0.9	0.81

51. $(1.2)^2 = 1.44$

53. $(-1.3)^2 = 1.69$

55. $-4.6(23.4 - 19.6) = -4.6(3.8) = -17.48$

57. $(-0.2)^2 + 2(7.1) = 0.04 + 14.2 = 14.24$

59. $(-0.7 - 0.5)(2.4 - 3.1) = -1.2(-0.7) = 0.84$

61. $(0.5 + 0.6)^2(-3.2) = (1.1)^2(-3.2) = 1.21(-3.2) = -3.872$

63. $|-2.6| \bullet |-7.2| = 2.6(7.2) = 18.72$

65. $(|-2.6 - 6.7|)^2 = (|-9.3|)^2 = (9.3)^2 = 86.49$

67. $3.14 + 2(1.2 - (-6.7)) = 3.14 + 2(7.9) = 18.94$

69. $-0.4 + 0.5(100)(-0.4)^2 = -0.4 + 8 = 7.6$

71. $10\left|(-1.1)^2 - 2.2^2\right| = 10|-3.63| = 36.3$

Applications

73. CONCERT SEATING

a.

Ticket	Price	Number	Receipts
Floor	\$12.50	1,000	$12.50(1{,}000) = \$12{,}500$
Balcony	\$15.75	100	$15.75(100) = \$1{,}575$

b. The total receipts were $12{,}500 + 1{,}575 = \$14{,}075$.

75. STORM DAMAGE

During the three-week period the house fell $0.57 + 2(0.09) = 0.75$ inches.

77. WEIGHTLIFTING

There are $2(45.5 + 20.5 + 2.2) = 136.4$ pounds on this bar.

79. BAKERY SUPPLIES

Type	Price	Pounds	Cost
Almonds	\$3.25	16	$16(3.25) = \$52$
Walnuts	\$2.10	25	$25(2.10) = \$52.50$
Peanuts	\$1.85	x	$\$1.85x$

81. SWIMMING POOL CONSTRUCTION

The perimeter of the pool is $2(50) + 2(30.3) = 160.6$ meters, which is the amount of coping needed.

83. BIOLOGY

The three dimensions from top to bottom are

$34(0.000000004) = 0.000000136$ inch;

$3.4(0.000000004) = 0.0000000136$ inch;

$10(0.000000004) = 0.00000004$ inch.

85. $(-9.0089 + 10.0087)(15.3) = 15.29694$

87. $(18.18 + 6.61)^2 + (5 - 9.09)^2 = 631.2722$

89. ELECTRIC BILL

To the nearest cent the cost would be $\$0.14277(719) = \$102.65$.

Writing

91. Answers may vary.
93. Answers may vary.

Review

95.
$$\begin{aligned}\frac{x}{2}-\frac{x}{3}&=-2\\6\left(\frac{x}{2}-\frac{x}{3}\right)&=6(-2)\\3x-2x&=-12\\x&=-12\end{aligned}$$

97. The absolute value of negative three.

99. $-\frac{8}{8}=-1$

Section 5.4 Division with Decimals

Vocabulary

1. In the division $2.5\overline{)4.075} = 1.63$, the decimal 4.075 is called the **dividend**, the decimal 2.5 is the **divisor**, and 1.63 is the **quotient**.
3. The **mean** of several values is the sum of those values divided by the number of values.
5. The **mode** of several values is the value that occurs most often.

Concepts

7. To divide by a decimal, move the decimal point of the divisor so that it becomes a **whole** number. The decimal point of the dividend is then moved the same number of places to the **right**. The decimal point in the quotient is written directly **above** the decimal point of the dividend.

9. $45 = 45.0 = 45.000$ is a true statement.

11. Moving the decimal point one place to the right is the same as multiplying by ten.

13. Check the result by $0.9(2.13) = 1.917$

15. This is the correct result since $2.07(4.6) = 9.522$.

17. The mean is $\frac{2.3+2.3+3.6+3.8+4.5}{5} = \frac{16.5}{5} = 3.3$, the median is 3.6 and the mode is 2.3.

Notation

19. The arrows indicated decimal shifts two places to the right or multiplying by 100.

Practice

21. $\begin{array}{r} 4.5 \\ 8\overline{)\ 36} \end{array}$

23. $-39 \div 4 = -9.75$

25. $49.6 \div 8 = 6.2$

27. $\begin{array}{r} 32.1 \\ 9\overline{)288.9} \end{array}$

29. $(-14.76) \div (-6) = 2.46$

31. $\frac{-55.02}{7} = -7.86$

33. $45\overline{)119.7}$ with quotient 2.66

35. $250.95 \div 35 = 7.17$

37. $41.6 \div 0.32 = 130$

39. $(-199.5) \div (-0.19) = 1,050$

41. $\dfrac{0.0102}{0.017} = 0.6$

43. $\dfrac{0.0186}{0.031} = 0.6$

45. $3\overline{)16} \approx 5.3$

47. $-5.714 \div 2.4 \approx -2.4$

49. $12.243 \div 0.9 \approx 13.60$

51. $0.04\overline{)0.03164} \approx 0.79$

53. $7.895 \div 100 = 0.07895$

55. $0.064 \div (-100) = -0.00064$

57. $1,000\overline{)34.8} = 0.0348$

59. $\dfrac{45.04}{10} = 4.504$

61. $\dfrac{-1.2-3.4}{3(1.6)} = \dfrac{-4.6}{4.8} \approx -0.96$

63. $\dfrac{40.7(-5.3)}{0.4-0.61} = \dfrac{-215.71}{-0.21} \approx 1,027.19$

65. $\dfrac{5(48.38-32)}{9} = \dfrac{81.9}{9} = 9.1$

67. $\dfrac{6.7-(0.3)^2+1.6}{0.3^3} = \dfrac{8.21}{0.027} \approx 304.07$

Applications

69. BUTCHER SHOP

There will be $\frac{14}{0.05} = 280$ slices.

71. HIKING

The hiker will hike for $\frac{27.5}{2.5} = 11$ hours, arriving at 6:00 p.m.

73. SPRAY BOTTLE

In an 8.5 ounce bottle there would be $\frac{8.5}{0.015} \approx 566.67$ or 567 sprays.

75. HOURLY PAY

In 1988 the average pay was $\frac{322.02}{34.7} = \$9.28$ per hour and in 1998 the average pay was

$\frac{441.84}{34.6} = \$12.77$ per hour.

77. OIL WELL

The total they must drill is $0.68 + 0.36 + 0.44 = 1.48$ miles.
To do this in four weeks they need to drill $1.48 \div 4 = 0.37$ miles per week.

79. OCTUPLETS

The mean birth weight was $\frac{24+27+28+26+11.2+17.5+28.5+18}{8} = 22.525$ ounces.

The median birth weight was $\frac{24+26}{2} = 25$ ounces.

81. COMPARISON SHOPPING

The mean was

$$\frac{3.89+3.97+3.98+3.99+4.09+4.19+4.19+4.24+4.29+4.29+4.29+4.39}{12} = \$4.15.$$

The median \$4.19 and the mode is \$4.29.

83. $\frac{8.6+7.99+(4.05)^2}{4.56} = \frac{32.9925}{4.56} \approx 7.24$

85. $\left(\frac{45.9098}{-234.12}\right)^2 - 4 \approx 0.0384533 - 4 \approx -3.96$

Writing

87. Answers may vary.
89. Answers may vary.

Review

91. $\dfrac{\frac{7}{8}}{\frac{3}{4}} = \dfrac{7}{8} \bullet \dfrac{4}{3} = \dfrac{7}{6}$

93. The integers are $\{\ldots,-3,-2,-1,0,1,2,3,\ldots\}$.

95. $-\dfrac{3}{4}A = -9$

$$A = -\frac{4}{3}(-9)$$

$$A = 12$$

97. $5x - 6(x-1) - (-x) = 5x - 6x + 6 + x = 6$

Estimation Study Set

1. The deluxe model is approximately $240 more expensive.
3. There is approximately 2 cubic feet less.
5. For this amount she can purchase approximately 30 standard models.
7. To split the cost three ways each will pay approximately $330.
9. There is about $520 left to finance.
11. This is not reasonable as the sum is about 26 more than 345.
13. This seems reasonable.
15. This seems reasonable.
17. This is not reasonable as 53 multiplied by 5 is 265.

Section 5.5 Fractions and Decimals

Vocabulary

1. The decimal form of the fraction $\frac{1}{3}$ is a **repeating** decimal, which is written $0.\bar{3}$ or 0.333....

3. The **decimal** equivalent of $\frac{1}{16}$ is 0.0625.

Concepts

5. $7 \div 8$ is indicated.

7. When rounding 0.272727... to the nearest hundredth the result is smaller.

9. From left to right $\{-3.8\bar{3}, -0.75, 0.\bar{6}, 1\frac{3}{4}\}$

-7 -6 -5 -4 -3 -2 -1 0 1 2 3 4 5 6 7

11. a. $\frac{1}{3} = 0.3$ is false

 b. $\frac{3}{4} = 0.75$ is true

 c. $20\frac{1}{2} = 20.5$ is true

 d. $\frac{1}{16} = 0.1\bar{6}$ is false

Notation

13. a. The remainder will never be zero.

 b. The decimal equivalent of $\frac{5}{6}$ is a repeating decimal.

Practice

15. $\frac{1}{2} = 0.5$

17. $-\frac{5}{8} = -0.625$

19. $\frac{9}{16} = 0.5625$

21. $-\frac{17}{32} = -0.53125$

23. $\frac{11}{20} = 0.55$

25. $\frac{31}{40} = 0.775$

27. $-\frac{3}{200} = -0.015$

29. $\frac{1}{500} = 0.002$

31. $\frac{2}{3} = 0.\overline{6}$

33. $\frac{5}{11} = 0.\overline{45}$

35. $-\frac{7}{12} = -0.58\overline{3}$

37. $\frac{1}{30} = 0.0\overline{3}$

39. $\frac{7}{30} \approx 0.23$

41. $\frac{17}{45} \approx 0.38$

43. $\frac{5}{33} \approx 0.152$

45. $\frac{10}{27} \approx 0.370$

47. $\frac{4}{3} \approx 1.33$

49. $-\frac{34}{11} \approx -3.09$

51. $3\frac{3}{4} = 3.75$

53. $-8\frac{2}{3} \approx -8.67$

55. $12\frac{11}{16} = 12.6875$

57. $203\frac{11}{15} \approx 203.73$

59. $\frac{7}{8} = 0.875 < 0.895$

61. $-\frac{11}{20} = -0.55 < -0.\bar{4}$

63. $\frac{1}{9} + \frac{3}{10} = \frac{10}{90} + \frac{27}{90} = \frac{37}{90}$

65. $\frac{9}{10} - \frac{7}{12} = \frac{54}{60} - \frac{35}{60} = \frac{19}{60}$

67. $\frac{5}{11}\left(\frac{3}{10}\right) = \frac{3}{22}$

69. $\frac{1}{3}\left(-\frac{1}{15}\right)\left(\frac{1}{2}\right) = -\frac{1}{45}\left(\frac{1}{2}\right) = -\frac{1}{90}$

71. $0.24 + \frac{1}{3} \approx 0.24 + 0.\bar{3} \approx 0.57$

73. $5.69 - \frac{5}{12} \approx 5.69 - 0.41\bar{6} \approx 5.27$

75. $(3.5 + 6.7)\left(-\frac{1}{4}\right) = 10.2(-0.25) = -2.55$

77. $\left(\frac{1}{5}\right)^2 (1.7) = 0.04(1.7) = 0.068$

79. $7.5 - (0.78)\left(\frac{1}{2}\right) = 7.5 - 0.39 = 7.11$

81. $\frac{3}{8}(-3.2) + (4.5)\left(-\frac{1}{9}\right) = -1.2 - 0.5 = -1.7$

83. $\frac{4}{3}(3.14)(3)^3 = 4(3.14)(9) = 113.04$

85. $\frac{23}{101} = 0.\overline{2277}$

87. $\frac{1,736}{50} = 34.72$

Applications

89. DRAFTING

$\frac{1}{16} = 0.0625$; $\frac{6}{16} = 0.375$; $\frac{9}{16} = 0.5625$; $\frac{15}{16} = 0.9375$

91. GARDENING

$0.065 = \frac{65}{1,000} = \frac{13}{200}; \frac{3}{40} = \frac{15}{200}$ thus the $\frac{3}{40}$ inch line is thicker.

Or $\frac{3}{40} = 0.075 > 0.65$ so the $\frac{3}{40}$ inch line is thicker.

93. HORSE RACING

$23^2 = 23\frac{2}{5} = 23.4$ sec; $23^4 = 23\frac{4}{5} = 23.8$ sec; $24^1 = 24\frac{1}{5} = 24.2$ sec; $32^3 = 32\frac{3}{5} = 32.6$ sec

95. WINDOW REPLACEMENT

The area of the window is found by multiplying the area of one triangular panel, $A = \frac{1}{2}(6)(5.2) = 15.6 \text{ in}^2$, by six since there are six panels, $6(15.6) = 93.6 \text{ in}^2$

Writing

97. Answers may vary.
99. Answers may vary.

Review

101. $-2 + (-3) + 10 + (-6) = -5 + 4 = -1$

103. $3T - 4T + 2(-4t) = -T - 8t$

105. $4x^2 + 2x^2 = 6x^2$

Section 5.6 Solving Equations Containing Decimals

Vocabulary

1. To **solve** an equation, we isolate the variable on one side of the equals sign.
3. In the term $5.65t$, the number 5.65 is called the **coefficient**.

Concepts

5. $2.1(1.7) - 6.3 \stackrel{?}{=} -2.73$

   $3.57 - 6.3 \stackrel{?}{=} -2.73$

   $-2.73 = -2.73$

7. Simplify applies to the expression, $7.8x + 9.1 + 12.4$.

9. a. \$0.25
   b. \$0.01
   c. \$2.50
   d. \$0.99

11. The distributive property of multiplication over subtraction is demonstrated.

Notation

13.
$$0.6s - 2.3 = -1.82$$
$$0.6s - 2.3 + 2.3 = -1.82 + 2.3$$
$$0.6s = 0.48$$
$$\frac{0.6s}{0.6} = \frac{0.48}{0.6}$$
$$s = 0.8$$

Practice

15. $8.7x + 1.4x = 10.1x$

17. $0.05h - 0.03h = 0.02h$

19. $3.1r - 5.5r - 1.3r = -3.7r$

21. $3.2 - 8.78x + 9.1 = 12.3 - 8.78x$

23. $5.6x - 8.3 - 6.1x + 12.2 = -0.5x + 3.9$

25. $0.05(100 - x) + 0.04x = 5 - 0.05x + 0.04x = 5 - 0.01x$

27. $$\begin{aligned} x+8.1&=9.8 \\ x+8.1-8.1&=9.8-8.1 \\ x&=1.7 \end{aligned}$$

29. $$\begin{aligned} 7.08&=t-0.03 \\ 7.08+0.03&=t-0.03+0.03 \\ 7.11&=t \end{aligned}$$

31. $$\begin{aligned} -5.6+h&=-17.1 \\ -5.6+5.6+h&=-17.1+5.6 \\ h&=-11.5 \end{aligned}$$

33. $$\begin{aligned} 7.75&=t-(-7.85) \\ 7.75&=t+7.85 \\ 7.75-7.85&=t+7.85-7.85 \\ -0.1&=t \end{aligned}$$

35. $$\begin{aligned} 2x&=-8.72 \\ x&=\frac{-8.72}{2} \\ x&=-4.36 \end{aligned}$$

37. $$\begin{aligned} -3.51&=-2.7x \\ \frac{-3.51}{-2.7}&=x \\ 1.3&=x \end{aligned}$$

39. $$\begin{aligned} \frac{x}{2.04}&=-4 \\ x&=-4(2.04) \\ x&=-8.16 \end{aligned}$$

41. $$\begin{aligned} \frac{-x}{5.1}&=-4.4 \\ x&=-5.1(-4.4) \\ x&=22.44 \end{aligned}$$

43. $$\begin{aligned} \frac{1}{3}x&=-7.06 \\ x&=3(-7.06) \\ x&=-21.18 \end{aligned}$$

45. $\frac{x}{100} = 0.004$

$x = 100(0.004)$

$x = 0.4$

47. $2x + 7.8 = 3.4$

$2x = -4.4$

$x = -2.2$

49. $-0.8 = 5y + 9.2$

$-10 = 5y$

$-2 = y$

51. $0.3x - 2.1 = 7.2$

$0.3x = 9.3$

$x = 31$

53. $-1.5b + 2.7 = 1.2$

$-1.5b = -1.5$

$b = 1$

55. $0.4a - 6 + 0.5a = -5.73$

$0.9a = 0.27$

$a = 0.3$

57. $2(t - 4.3) + 1.2 = -6.2$

$2t - 8.6 + 1.2 = -6.2$

$2t - 7.4 = -6.2$

$2t = 1.2$

$t = 0.6$

59. $1.2x - 1.3 = 2.4x + 0.02$

$-1.32 = 1.2x$

$-1.1 = x$

61. $53.7t - 10.1 = 46.3t + 4.7$

$7.4t = 14.8$

$t = 2$

63. $2.1x - 4.6 = 7.3x - 11.36$

$6.76 = 5.2x$

$1.3 = x$

65. $0.06x + 0.09(100 - x) = 8.85$

$$0.06x + 9 - 0.09x = 8.85$$
$$-0.03x = -0.15$$
$$x = 5$$

Applications

67. PETITION DRIVE

Analyze the problem

- Her base pay is 15 dollars a day.
- She makes 30 cents for each signature.
- She wants to make 60 dollars a day.
- Find the number of signatures she needs to get.

Form an equation

Let x = the number of signatures she needs to collect.
We need to work in terms of the same units, so we write 30 cents as $0.30.
If we multiply the pay per signature by the number of signatures, we get the money she makes just from collecting signatures. Therefore, $0.30x =$ total amount (in dollars) made from collecting signatures.
Base pay plus 0.30 times the number of signatures is 60.
$15 + 0.30x = 60$

Solve the equation

$$15 + 0.30x = 60$$
$$0.30x = 45$$
$$x = 150$$

State the conclusion

She needs to collect 150 signatures to make $60.

Check the result

If she collects 150 signatures, she will make $0.30(150) = \$45$ from signatures. If we add the $15, we get $60. The answer checks.

69. DISASTER RELIEF

Let d represent the amount of relief to request at the federal level, then

$$6.8 + 12.5 + d = 27.9$$
$$19.3 + d = 27.9$$
$$d = \$8.6$$

The county should request $8.6 million from the federal government.

71. GPA

Let g represent her grade point before the decline.

$$g-0.18=3.09$$
$$g=3.27$$

Before the drop in grade point her average was 3.27.

73. POINTS PER GAME

Let j represent her scoring average as a junior.

$$2j=21.4$$
$$j=\frac{21.4}{2}$$
$$j=10.7$$

As a junior her average was 10.7 points per game.

75. FUEL EFFICIENCY

Let s represent the fuel average in 1960.

$$s-0.4+1.3+3.1+0.3=16.7$$
$$s+4.3=16.7$$
$$s=12.4$$

In 1960 the average miles per gallon was 12.4.

77. CALLIGRAPHY

Let w represent words.

$$20+0.15w=50$$
$$0.15w=30$$
$$w=200$$

The maximum number of words is 200.

Writing

79. Answers may vary.

Review

81. $-\frac{2}{3}+\frac{3}{4}=-\frac{8}{12}+\frac{9}{12}=\frac{1}{12}$

83. $\left(-\frac{1}{2}\right)^3-(-1)^3=-\frac{1}{8}+1=\frac{7}{8}$

85. $\frac{-3-3}{-3+4}=\frac{-6}{1}=-6$

Section 5.7 Square Roots

Vocabulary

1. When we find what number is squared to obtain a given number, we are finding the square **root** of the given number.
3. The symbol $\sqrt{\ }$ is called a **radical** sign. It indicates that we are to find a **positive** square root.
5. In $\sqrt{26}$, 26 is called the **radicand**.

Concepts

7. The square of 5 is 25, because $5^2 = 25$.

9. The two square roots of 49 are 7 and –7 because $7^2 = 49$ and $(-7)^2 = 49$.

11. Since $\left(\frac{3}{4}\right)^2 = \frac{9}{16}$, we know that $\sqrt{\frac{9}{16}} = \frac{3}{4}$.

13. From smallest to largest, $\sqrt{6}, \sqrt{11}, \sqrt{23}, \sqrt{27}$.

15. a. $\sqrt{1} = 1$
    b. $\sqrt{0} = 0$

17. a. $\sqrt{6} \approx 2.4$.
    b. $2.4^2 = 5.76$
    c. $6 - 5.76 = 0.24$

19. From left to right, $-\sqrt{5}, \sqrt{9}$.

-5 -4 -3 -2 -1 0 1 2 3 4 5

21. a. $\sqrt{19}$ is between 4 and 5.
    b. $\sqrt{87}$ is between 9 and 10.

Notation

23. $-\sqrt{49} + \sqrt{64} = -7 + 8$
    $= 1$

Practice

25. $\sqrt{16} = 4$

27. $-\sqrt{121} = -11$

29. $-\sqrt{0.49} = -0.7$

31. $\sqrt{0.25} = 0.5$

33. $\sqrt{0.09} = 0.3$

35. $-\sqrt{\frac{1}{81}} = -\frac{1}{9}$

37. $-\sqrt{\frac{16}{9}} = -\frac{4}{3}$

39. $\sqrt{\frac{4}{25}} = \frac{2}{5}$

41. $5\sqrt{36} + 1 = 5(6) + 1 = 31$

43. $-4\sqrt{36} + 2\sqrt{4} = -4(6) + 2(2) = -20$

45. $\sqrt{\frac{1}{16}} - \sqrt{\frac{9}{25}} = \frac{1}{4} - \frac{3}{5} = \frac{5}{20} - \frac{12}{20} = -\frac{7}{20}$

47. $5(\sqrt{49})(-2) = 5(7)(-2) = -70$

49. $\sqrt{0.04} + 2.36 = 0.2 + 2.36 = 2.56$

51. $-3\sqrt{1.44} = -3(1.2) = -3.6$

53.

Number	Square Root
1	1.000
2	1.414
3	1.732
4	2.000
5	2.236
6	2.449
7	2.646
8	2.828
9	3.000
10	3.162

55. $\sqrt{1,369} = 37$

57. $\sqrt{3,721} = 61$

59. $\sqrt{15} \approx 3.87$

61. $\sqrt{66} \approx 8.12$

63. $\sqrt{24.05} \approx 4.904$

65. $-\sqrt{11.1} \approx -3.332$

67. $\sqrt{24,000,201} = 4,899$

69. $-\sqrt{0.00111} \approx -0.0333$

Applications

71. CARPENTRY
    a. The slanted part can be considered the hypotenuse of a right triangle so its length is $\sqrt{3^2 + 4^2} = \sqrt{25} = 5$ feet.
    b. The slanted part can be considered the hypotenuse of a right triangle so its length is $\sqrt{6^2 + 8^2} = \sqrt{100} = 10$ feet.

73. BASEBALL DIAMOND
From home plate to second base is $\sqrt{16,200} \approx 127.3$ feet.

75. BIG-SCREEN TV
The screen size is $\sqrt{1,681} = 41$ inches.

Writing

77. Answers may vary.
79. Answers may vary.
81. Answers may vary.

Review

83. To isolate the variable we must first undo the subtraction then undo the multiplication.

85. $5(-2)^2 - \dfrac{16}{4} = 5(4) - 4 = 16$

87. The whole numbers $\{0,1,2,3,4,5,...\}$.

89. $8+\frac{a}{5}=14$

$\frac{a}{5}=6$

$a=5(6)$

$a=30$

The Real Numbers

1. From left to right $\{-4, -2\frac{1}{2}, -0.1, \frac{99}{100}, 1.\overline{3}, \frac{13}{4}, \sqrt{17}\}$.

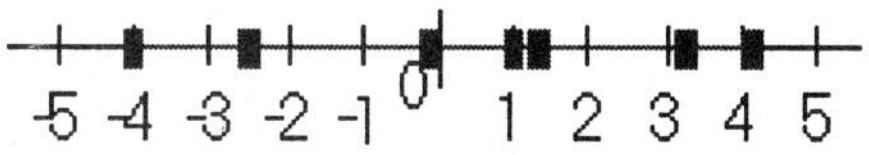

-5 -4 -3 -2 -1 0 1 2 3 4 5

3. Natural Numbers $\{1, 2, 3, 4, 5, \ldots\}$.
5. Integers $\{\ldots, -3, -2, -1, 0, 1, 2, 3, 4, 5, \ldots\}$.
7. Irrational numbers are non-repeating, non-terminating decimals that cannot be expressed in the form $\frac{a}{b}$.
9. Every fraction can be written as a terminating decimal is **false.**
11. Some irrational numbers are integers is **false.**
13. No numbers are both rational and irrational numbers is **true.**
15. The set of whole numbers is a subset of the irrational numbers is **false.**
17. Every natural number is an integer is **true.**
19. There is no real number such that when it is squared the result is a negative number.

Chapter 5 Key Concept

1. a. $-2x+3+7x-11=5x-8$

   b. $-2x+3+7x-11=7$
   $$5x-8=7$$
   $$5x=15$$
   $$x=3$$

3. a. $3(0.2y-1.6)+0.6y=0.6y-4.8+0.6y=1.2y-4.8$

   b. $3(0.2y-1.6)+0.6y=-6.6$
   $$1.2y-4.8=-6.6$$
   $$1.2y=-1.8$$
   $$y=-1.5$$

Chapter 5 Review

Section 5.1 An Introduction to Decimals

1. As a decimal the shaded region is 0.67, as a fraction the shaded region is $\frac{67}{100}$.

3. $16.4523 = 10 + 6 + \frac{4}{10} + \frac{5}{100} + \frac{2}{1,000} + \frac{3}{10,000}$.

5. From left to right $\{-2.7, -0.8, 1.55\}$.

-5 -4 -3 -2 -1 0 1 2 3 4 5

7. This is a true statement.

9. a. $4.578 \approx 4.58$
   b. $3,706.0895 \approx 3,706.090$
   c. $-0.0614 \approx -0.1$
   d. $88.12 \approx 88.1$

Section 5.2 Addition and Subtraction with Decimals

11. a. $-16.1 + 8.4 = -7.7$
    b. $-4.8 - (-7.9) = 3.1$
    c. $-3.55 + (-1.25) = -4.8$
    d. $-15.1 - 13.99 = -29.09$
    e. $-8.8 + (-7.3 - 9.5) = -8.8 + (-16.8) = -25.6$
    f. $(5 - 0.096) - (-0.035) = 4.904 + 0.035 = 4.939$

13. MICROWAVE OVEN
    The window is $13.4 - (2.5 + 2.75) = 13.4 - 5.25 = 8.15$ inches tall.

Section 5.3 Multiplication with Decimals

15. a. $1,000(90.1452) = 90,145.2$
    b. $(-10)(-2.897)(100) = 28.97(100) = 2,897$

17. a. $(0.6 + 0.7)^2 - 12.3 = 1.3^2 - 12.3 = -10.61$
    b. $3(7.8) + 2(1.1)^2 = 23.4 + 2(1.21) = 23.4 + 2.42 = 25.82$

19. WORD PROCESSOR

The height of the printed portion will be $11-(1.0+0.6)=9.4$ inches.

The width of the printed portion will be $8.5-(0.5+0.7)=7.3$ inches.

The area of the printed portion will be $A=9.4(7.3)=68.62$ square inches.

Section 5.4 Division with Decimals

21. a. $12\overline{)\,15\,}$ with quotient 1.25

b. $-41.8\div 4=-10.45$

c. $\dfrac{-29.67}{-23}=1.29$

d. $24.618\div 6=4.103$

23. a. $78.98\div 6.1\approx 12.9$

b. $\dfrac{-5.338}{0.008}\approx -667.3$

25. THANKSGIVING DINNER

The cost per person for dinner was $\dfrac{\$41.70}{5}=\$8.34$.

27. $\dfrac{(1.4)^2+2(4.6)}{0.5+0.3}=\dfrac{1.96+9.2}{0.8}=13.95$

29. TELESCOPE

This adjustment requires $\dfrac{0.2375}{0.025}=9.5$ revolutions.

31. TOBACCO SETTLEMENT

To find the median place the dollar values in order, in this case the order is smallest to largest,

State	$ Billions
Vermont	0.81
New Hampshire	1.3
Rhode Island	1.4
Maine	1.5
Connecticut	3.63
Massachusetts	8

The median is $\frac{1.4+1.5}{2}=\$1.45$ billion.

Section 5.5 Fractions and Decimals

33. a. $\frac{6}{11}=0.\overline{54}$

b. $-\frac{2}{3}=-0.\overline{6}$

35. a. $\frac{13}{25}=0.52>0.499$

b. $-0.\overline{26}>-\frac{4}{15}=-0.2\overline{7}$

37. a. $\frac{1}{3}+0.4=\frac{1}{3}+\frac{4}{10}=\frac{1}{3}+\frac{2}{5}=\frac{5}{15}+\frac{6}{15}=\frac{11}{15}$

b. $\frac{4}{5}(-7.8)=0.8(-7.8)=-6.24$

c. $\frac{1}{2}(9.7+8.9)(10)=5(18.6)=93$

d. $\frac{1}{3}(3.14)(3)^2(4.2)=(3.14)(3)(4.2)=39.564$

39. ROADSIDE EMERGENCY

The area of this reflector is $A=\frac{1}{2}(10.9)(6.4)=34.88$ square inches.

Section 5.6 Solving Equations Containing Decimals

41. $-1.3+1.2(-1.1)\overset{?}{=}2.4(-1.1)+0.02$

    $-2.62=-2.62$

    So $r=-1.1$ is a solution.

43. a. $1.7y+1.24=-1.4y-0.62$

    $3.1y=-1.86$

    $y=-0.6$

    b. $0.05(1{,}000-x)+0.9x=60.2$

    $50-0.05x+0.9x=60.2$

    $50+0.85x=60.2$

    $0.85x=10.2$

    $x=12$

Section 5.7 Square Roots

45. Two square roots of 64 are 8 and –8 because $8^2=64$ and $(-8)^2=64$.

47. When graphed on a number line $\sqrt{83}$ would lie between 9 and 10.

49. From left to right $\{-\sqrt{2},\sqrt{0},\sqrt{3}\}$

-5 -4 -3 -2 -1 0 1 2 3 4 5

51. a. $\sqrt{19}\approx 4.36$

    b. $\sqrt{59}\approx 7.68$

Chapter 5 Test

1. The amount of shaded squares is $\frac{79}{100} = 0.79$.

3. $0.271 = \frac{271}{1,000}$

5. SKATING RECEIPTS
   For the two days the total income was $30.25 + 62.25 + 40.50 + 75.75 = \$208.75$.

7. EARTHQUAKE FAULT LINE
   The total drop was $0.83 + 0.19 = 1.02$ inches.

9. NEW YORK CITY
   The area of central park is $2.5(0.5) = 1.25 \text{ mi}^2$.

11. $4.1 - (3.2)(0.4)^2 = 4.1 - (3.2)(0.16) = 3.588$

13. $\frac{12.146}{-5.3} \approx -2.29$

15. The mean is $\frac{2.9 + 3.1 + 3.1 + 3.6 + 3.9 + 4.4 + 4.5}{7} \approx 3.6$.
    The median is 3.6 and the mode is 3.1.

17. From left to right $\left\{-\frac{4}{5}, \frac{3}{8}\right\}$, the decimal equivalents are $\{-0.8, 0.375\}$.

-3 -2 -1 0 1 2 3

19. $6.18s + 8.9 - 1.22s - 6.6 = 4.96s + 2.3$

21. a. $-2.4t = 16.8$
    $t = -7$

    b. $-0.008 + x = 6$
    $x = 6.008$

23. CHEMISTRY
    The weight of compound C is $1.86 + 2.09 + C = 4.37$
    $C = 0.42$ grams

25. From left to right $\{-\sqrt{5}, \sqrt{2}\}$.

-4 -3 -2 -1 0 1 2 3 4

27. a. $-6.78 > -6.79$

b. $\frac{3}{8} = 0.375 > 0.3$

c. $\sqrt{\frac{16}{81}} = \frac{4}{9} = \frac{36}{81} > \frac{16}{81}$

d. $0.\overline{45} > 0.45$

Chapters 1 – 5 Cumulative Review Exercises

1. THE EXECUTIVE BRANCH
   During a four year term the president earns $400,000(4) = \$1,600,000$ and the vice-president earns $181,400(4) = \$725,600$. The difference in earnings is $1,600,000 - 725,600 = \$874,400$.

3. $43\overline{)1,203}$ is 27 with 42 remaining.

5. The prime factorization of 220 is $2^2 \bullet 5 \bullet 11$.

7. The whole numbers are $\{0, 1, 2, 3, 4, 5, ...\}$.

9. Subtraction is the same as **adding** the opposite.

11. The equivalent multiplication statement is $-15 = -5 \bullet 3$.

13. $$\begin{aligned} 8 - 2d &= -5 - 5 \\ 8 - 2d &= -10 \\ -2d &= -18 \\ d &= 9 \end{aligned}$$

15. $|-7(5)| = |-35| = 35$

17. The length in feet of this chain is $3x$.

19. a. The length of the match is $(k+1)$ inches.
    b. The length of the key $(m-1)$ inches.

21. There are three terms in this expression.

23. $$\begin{aligned} -(5x - 4) + 6(2x - 7) &= -3 \\ -5x + 4 + 12x - 42 &= -3 \\ 7x - 38 &= -3 \\ 7x &= 35 \\ x &= 5 \end{aligned}$$

25. This represents equivalent fractions.

27. $\frac{3}{8} \bullet \frac{7}{16} = \frac{21}{128}$

29. $\frac{4}{m} + \frac{2}{7} = \frac{28}{7m} + \frac{2m}{7m} = \frac{28 + 2m}{7m}$

31. $76\frac{1}{6} - 49\frac{7}{8} = 76\frac{4}{24} - 49\frac{21}{24} = 75\frac{28}{24} - 49\frac{21}{24} = 26\frac{7}{24}$

33. $\frac{2}{3}y = -30$

$$y = -\frac{90}{2}$$

$$y = -45$$

35. KITE

The area of the kite is found by considering the kite as two triangles,

$$A = \frac{1}{2}\left(7\frac{1}{2}\right)(21)$$

$A = 78.75$ doubling this yields 157.5 in^2 which is the area of the kite.

37. GLASS

To the nearest thousandths 0.001 inch.

39. $-1.8(4.52) = -8.136$

41. $56.012(100) = 5,601.2$

43. $-9.1 - (-6.05 - 51) = -9.1 - (-57.05) = 47.95$

45. LITERATURE

$$C = \frac{5(451-32)}{9} = 232.\overline{7} \approx 232.8$$

47. $\frac{5}{12} = 0.41\overline{6}$

49. CONCESSIONAIRE

To make \$50 she must sell $22 + 0.35P = 50$

$$0.35P = 28$$

$$P = 80 \text{ bags}$$

Section 6.1 The Rectangular Coordinate System

Vocabulary

1. The pair of numbers $(2,5)$ is called an **ordered** pair.
3. Since the ordered pair $(1,3)$ is a solution of $x+y=4$, we say that $(1,3)$ **satisfies** the equation.
5. The rectangular coordinate system is sometimes called the **Cartesian** coordinate system.
7. The point where the *x*- and *y*-axes cross is called the **origin**.

Concepts

9. BURNING CALORIES

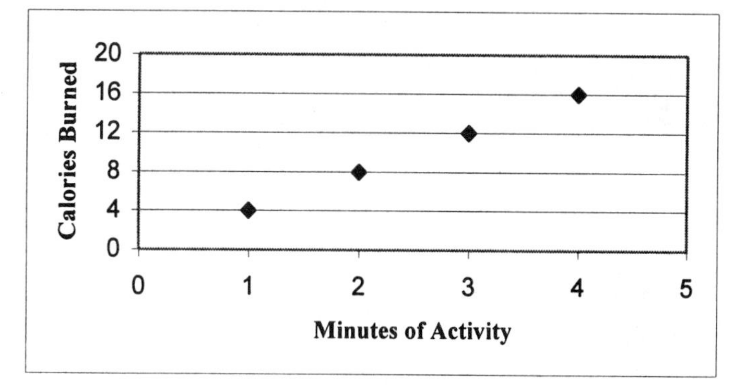

11. $(2,3)$ is not a solution to $2x+3y=14$ since $2(2)+3(3)=13\neq 14$.
13. To plot the point with coordinates $(3,-4)$, we start at the **origin** and move 3 units to the **right** and then move 4 units **down**.

Notation

15.
$$\begin{aligned} 4x+3y&=14\\ 4(2)+3y&=14\\ 8+3y&=14\\ 3y&=6\\ y&=2 \end{aligned}$$

17. These points are the same.

Practice

19. $3x + y = 12$
    a. $3(0) + y = 12 \rightarrow y = 12$
    b. $3x + 0 = 12 \rightarrow x = 4$
    c. $3(2) + y = 12 \rightarrow y = 6$

21. $2x + y = 8$
    a. $(0,8)$ since $2(0) + y = 8 \rightarrow y = 8$
    b. $(4,0)$ since $2x + 0 = 8 \rightarrow x = 4$
    c. $(3,2)$ since $2x + 2 = 8 \rightarrow x = 3$

23. $5x - 4y = 20$

x	y	(x, y)
0	-5	(0, -5)
4	0	(4, 0)
8	5	(8, 5)

25. 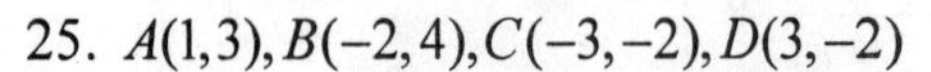
$A(1,3), B(-2,4), C(-3,-2), D(3,-2)$

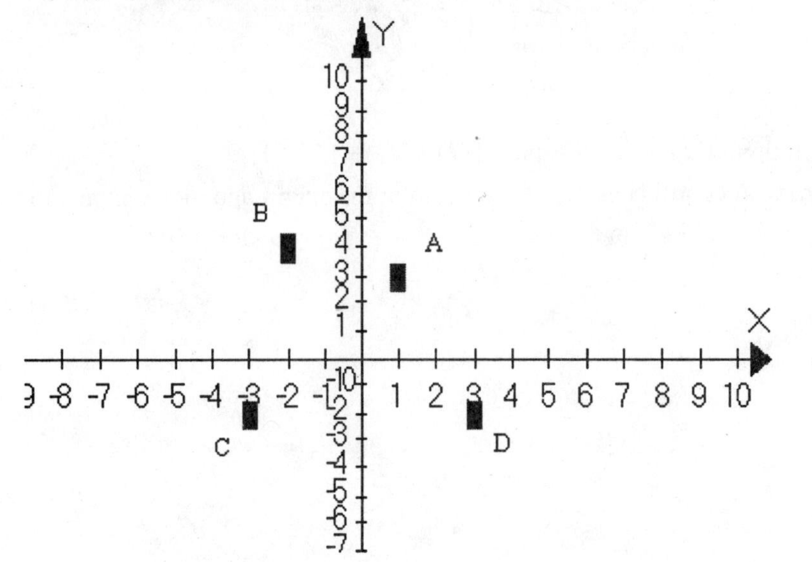

27. $A(-4,-3), B(1.5,1.5), C(-3.5,0), D(0,3.5)$

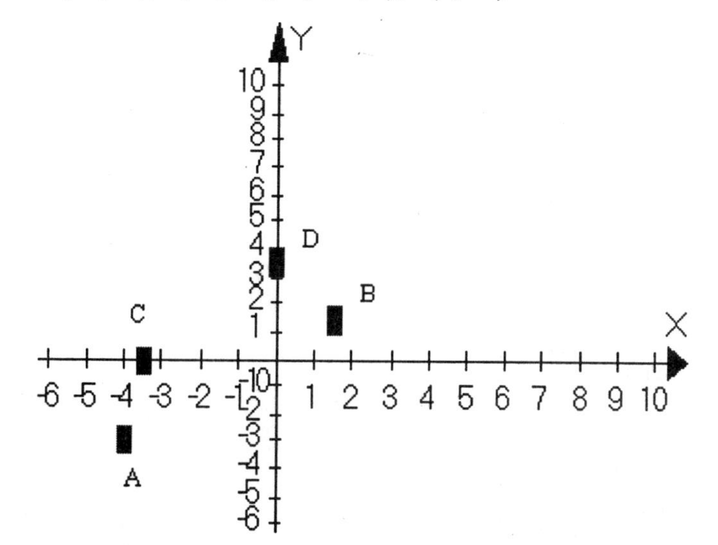

29. $A(2,4), B(-3,3), C(-2,-3), D(4,-3)$

31. $A(-3,-4), B(2.5,3.5), C(-2.5,0), D(2.5,0)$

Applications

33. ROAD MAPS
Rockford (5, B), Mount Carroll (1, C), Harvard (7, A), intersection (5, E)

35. EARTHQUAKE DAMAGE
   a. The coordinates of the epicenter are (2, -1).
   b. There was no damage at (4, 5).
   c. There was damage at (-1, -4).

37. THE GLOBE
The cities in order starting with the farthest east city are New Delhi, Kampala, Coats Land, Reykjavik, Buenos Aires, and Havana.

39. COOKING

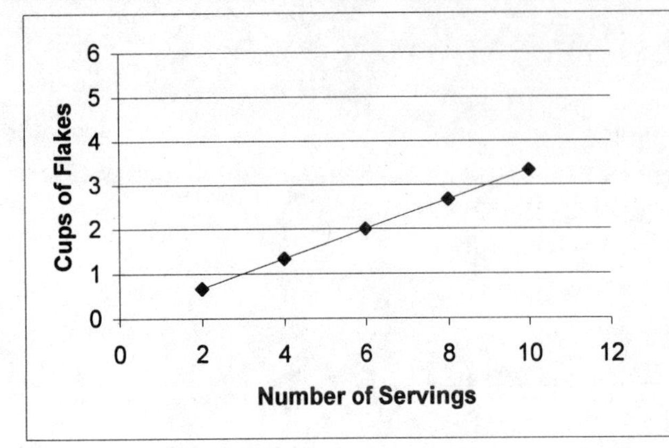

Writing

41. Answers may vary.
43. Answers may vary.

Review

45. $(-8-5)-3=-13-3=-16$

47. $(-4)^2-3^2=16-9=7$

49. $\frac{x}{3}+3=10$

$$\frac{x}{3}=7$$

$$x=21$$

51. $5-(7-x)=-5$

$$5-7+x=-5$$

$$x=-3$$

53. $(4^2)^4=65,536$

Section 6.2 Graphing Linear Equations

Vocabulary

1. The graph of linear equation is a **line**.
3. The point where the graph of a linear equation crosses the *x*-axis is called the ***x*-intercept**.
5. In the equation $y = 7x + 2$, *x* is called the **independent** variable.
7. The graph of the equation $y = 3$ is a **horizontal** line.

Concepts

9. a. The *y*-intercept is (0, 1).
   b. The *x*-intercept is (-2, 0).
   c. The line does pass through (4, 3).

11. Arrows are missing at each end of the line.

13. Answers may vary, but six solutions include $(20,20),(19,19),(0,0),(1.5,1.5),(2,2),(5,5)$.

Notation

15.
$$\begin{aligned} 2x - 4y &= 8 \\ 2(3) - 4y &= 8 \\ 6 - 4y &= 8 \\ -4y &= 2 \\ y &= -\frac{1}{2} \end{aligned}$$

Practice

17. $2x - 5y = 10$

x	y	(x, y)
5	0	(5, 0)
-5	-4	(-5, -4)
10	2	(10, 2)

19. $y = 2x - 3$

x	y	(x, y)
3	3	(3, 3)
-4	-11	(-4, -11)
6	9	(6, 9)

21. $x + y = 5$ has *y*-intercept (0, 5) and *x*-intercept (5, 0).

23. $4x + 5y = 20$ has y-intercept (0, 4) and x-intercept (5, 0).

25. $x + y = 5$

x	y
0	5
5	0
2	3

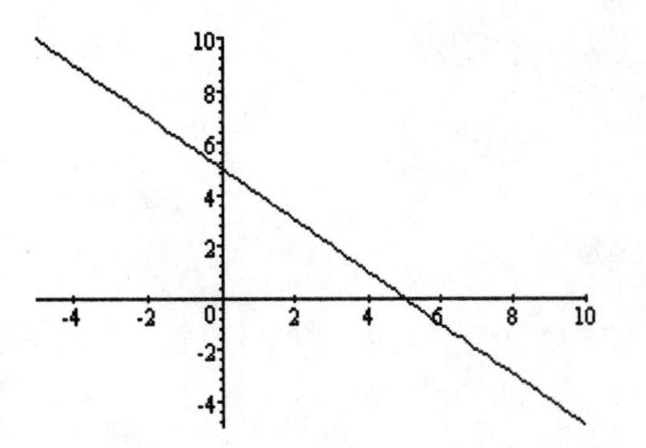

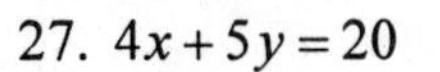
27. $4x + 5y = 20$

x	y
0	4
5	0
1	$\frac{16}{5}$

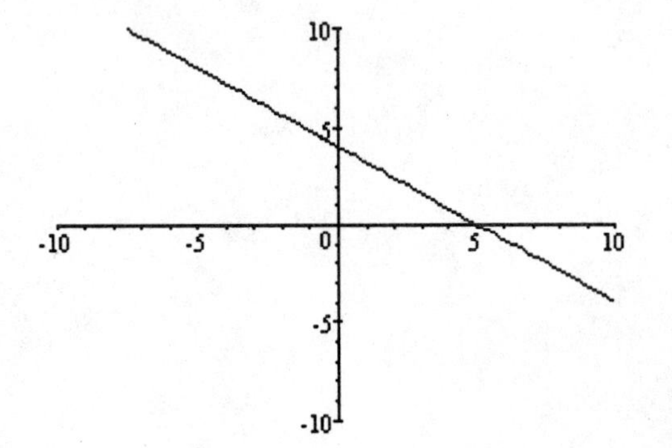

29. $x - 2y = -4$

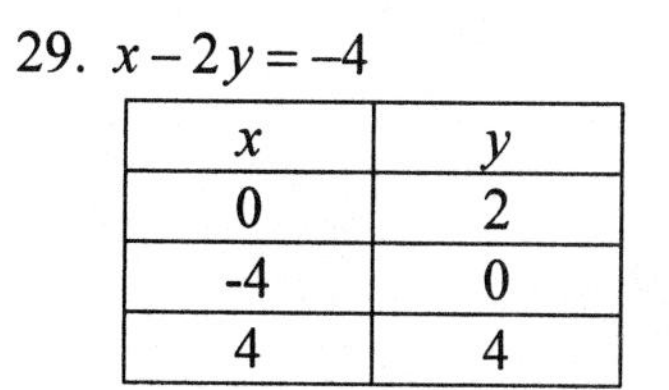

x	y
0	2
-4	0
4	4

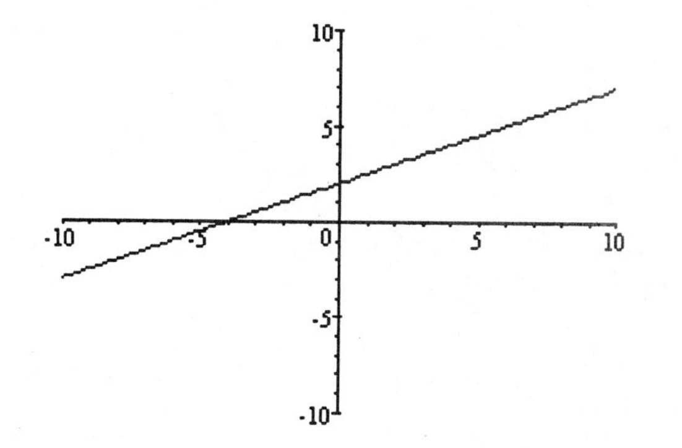

31. $y = 2x - 5$

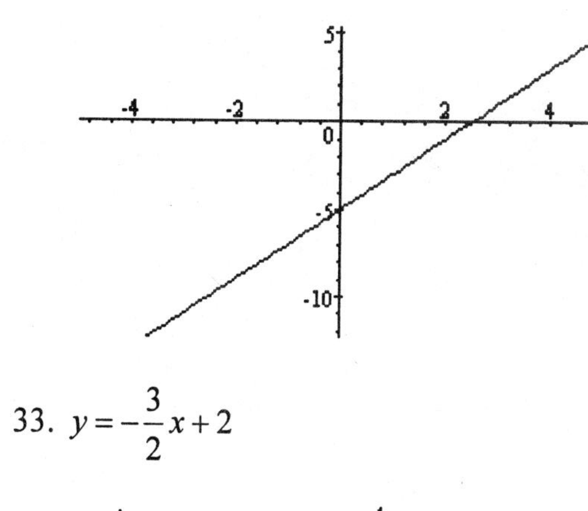

33. $y = -\dfrac{3}{2}x + 2$

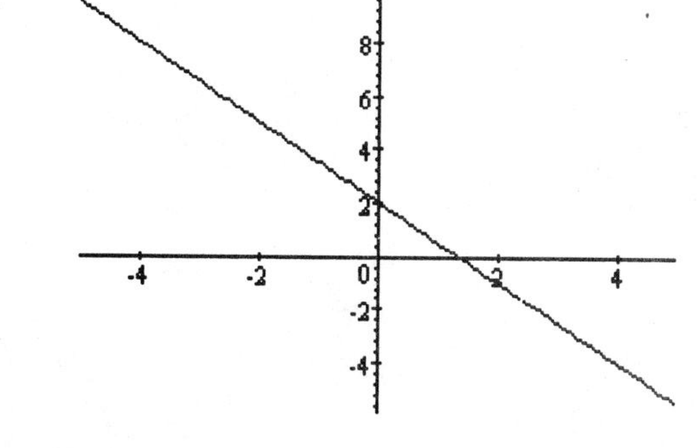

35. $y = 5$

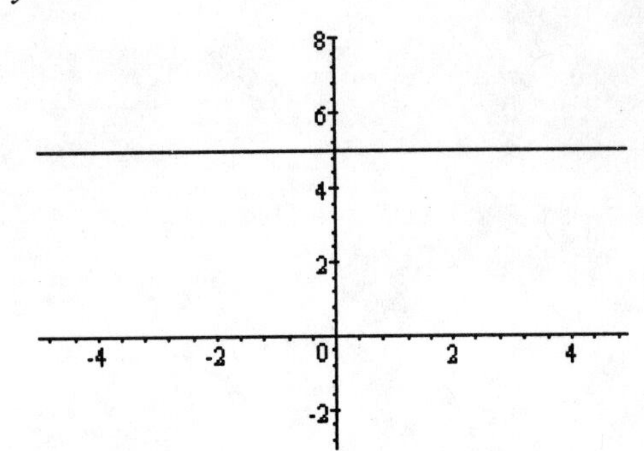

37. $x = 4$

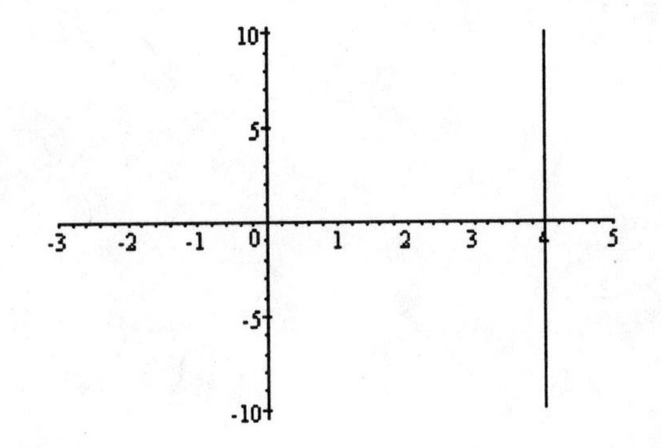

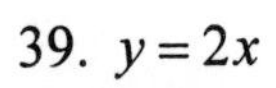
39. $y = 2x$

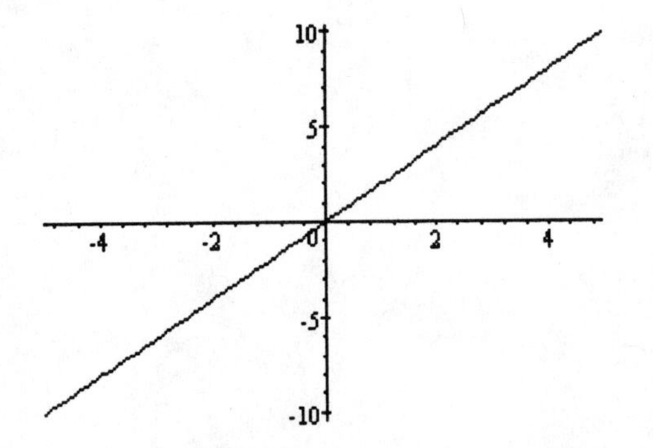

41. $y = \frac{x}{3}$

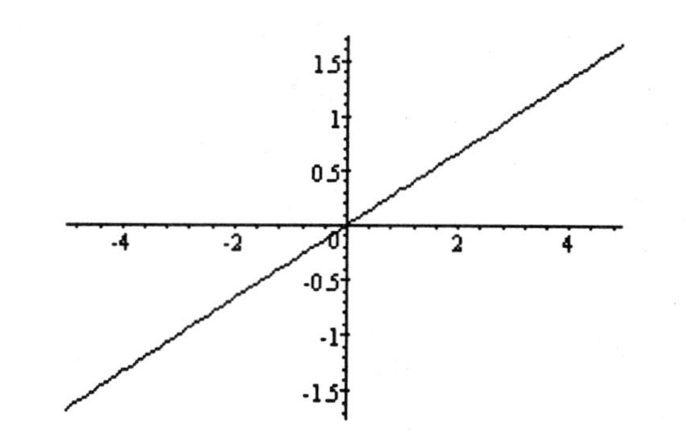

43. $y = 100x$

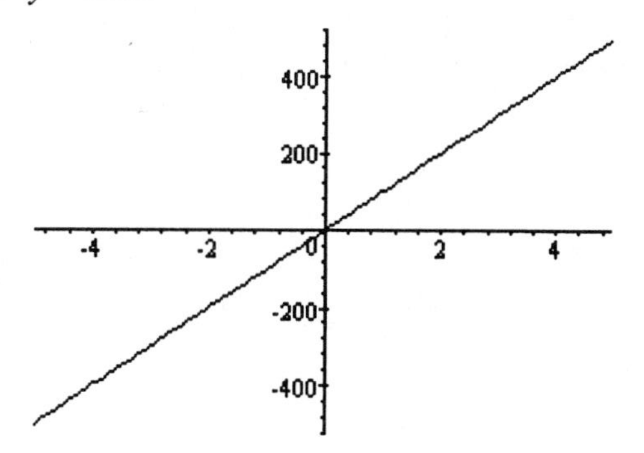

45. $y = -50x - 25$

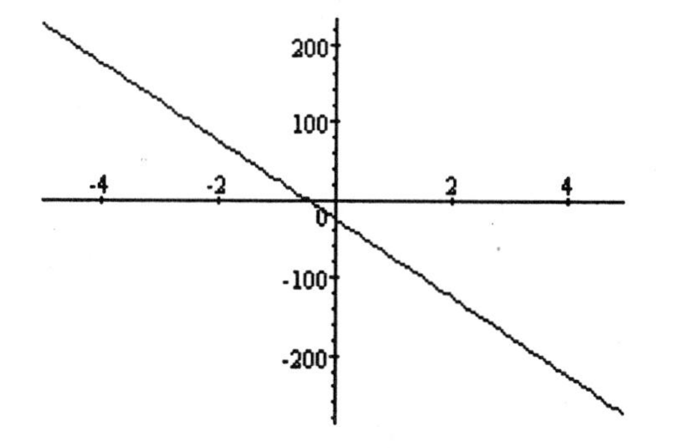

Applications

47. HOURLY WAGES

In three hours the student will earn approximately \$22.50.

x	y
2	15
4	30
6	45

$y = \frac{3}{2}x$

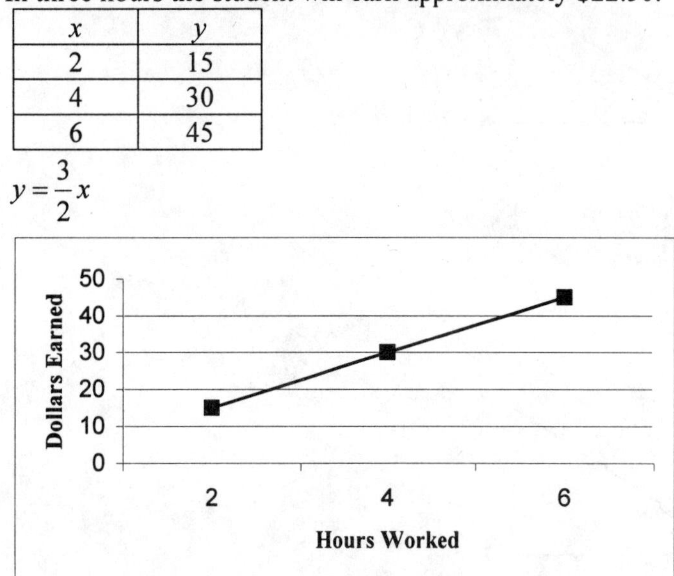

49. DISTANCE, RATE, AND TIME

t	$d = 2t$
1	2
2	4
3	6
4	8
5	10

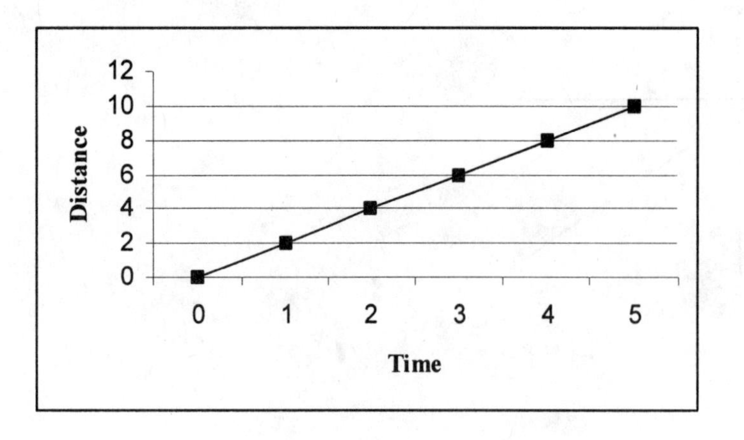

51. AIR TRAFFIC CONTROL

There is a possibility of a midair collision according to the graph. The point where the two graphs are equal is when $x = 2$ and $y = 2$, the point of collision.

$$-\frac{1}{2}x + 3 = \frac{2}{3}x + \frac{2}{3}$$
$$6\left(-\frac{1}{2}x + 3\right) = 6\left(\frac{2}{3}x + \frac{2}{3}\right)$$
$$-3x + 18 = 4x + 4$$
$$14 = 7x$$
$$2 = x$$

Writing

53. Answers may vary.
55. Answers may vary.
57. Answers may vary.

Review

59. $180 = 2^2 \bullet 3^2 \bullet 5$

61. $\dfrac{3(3-(-2))}{5(-2)+7} = \dfrac{3(5)}{-10+7} = \dfrac{15}{-3} = -5$

63. LIGHTENING

As a decimal in lowest terms $0.25 = \frac{1}{4}$.

Section 6.3 Multiplication Rules for Exponents

Vocabulary

1. In x^n, x is called the **base** and n is called the exponent.
3. $x^m \bullet x^n$ is the product of two exponential expressions with **like** bases.
5. $(2x)^n$ is a **product** raised to a power.

Concepts

7. a. $x \bullet x \bullet x \bullet x \bullet x \bullet x \bullet x = x^7$
   b. $x \bullet x \bullet y \bullet y \bullet y = x^2 y^3$
   c. $3 \bullet 3 \bullet 3 \bullet 3 \bullet a \bullet a \bullet b \bullet b \bullet b = 3^4 a^2 b^3$

9. Answers may vary, one solution is $x^2 x^6 = x^8$.
11. Answers may vary, one solution is $(y^5)^2 = y^{10}$.

13. a. $x^m x^n = x^{m+n}$
    b. $(x^m)^n = x^{mn}$
    c. $(ax)^n = a^n x^n$

15. a. $2^1 = 2$
    b. $(-10)^1 = -10$
    c. $x^1 = x$

17. a. $x \bullet x = x^2; x + x = 2x$
    b. $x \bullet x^2 = x^3; x + x^2 = x + x^2$
    c. $x^2 \bullet x^2 = x^4; x^2 + x^2 = 2x^2$

19. a. $4x \bullet x = 4x^2; 4x + x = 5x$
    b. $4x \bullet 3x = 12x^2; 4x + 3x = 7x$
    c. $4x^2 \bullet 3x = 12x^3; 4x^2 + 3x = 4x^2 + 3x$

21. $3^{2+1} = 3^3 = 27$

Notation

23. $x^5x^7 = x^{5+7}$
$= x^{12}$

25. $(2x^4)(8x^3) = (2 \bullet 8)(x^4x^3)$
$= 16x^{4+3}$
$= 16x^7$

Practice

27. $x^2x^3 = x^{2+3} = x^5$

29. $x^3x^7 = x^{3+7} = x^{10}$

31. $f^5f^8 = f^{5+8} = f^{13}$

33. $n^{24}n^8 = n^{24+8} = n^{32}$

35. $l^4l^5l = l^{4+5+1} = l^{10}$

37. $x^6x^3x^2 = x^{6+3+2} = x^{11}$

39. $2^4 \bullet 2^8 = 2^{4+8} = 2^{14}$

41. $5^6 \bullet 5^2 = 5^{6+2} = 5^8$

43. $2x^2 \bullet 4x = 8x^{2+1} = 8x^3$

45. $5t \bullet t^9 = 5t^{1+9} = 5t^{10}$

47. $-6x^3 \bullet 4x^2 = -24x^{3+2} = -24x^5$

49. $-x \bullet x^3 = -x^{1+3} = -x^4$

51. $6y(2y^3)(3y^4) = 36y^{1+3+4} = 36y^8$

53. $-2t^3(-4t^2)(-5t^5) = -40t^{3+2+5} = -40t^{10}$

55. $xy^2 \bullet x^2y = x^{1+2}y^{2+1} = x^3y^3$

57. $b^3c^2b^5c^6 = b^{3+5}c^{2+6} = b^8c^8$

59. $x^4y(xy) = x^{4+1}y^{1+1} = x^5y^2$

61. $a^2b \bullet b^3a^2 = a^{2+2}b^{1+3} = a^4b^4$

63. $x^5y \bullet y^6 = x^5y^{1+6} = x^5y^7$

65. $3x^2y^3 \bullet 6xy = 18x^{2+1}y^{3+1} = 18x^3y^4$

67. $xy^2 \bullet 16x^3 = 16x^{1+3}y^2 = 16x^4y^2$

69. $-6f^2t(4f^4t^3) = -24f^{2+4}t^{1+3} = -24f^6t^4$

71. $ab \bullet ba \bullet a^2b = a^{1+1+2}b^{1+1+1} = a^4b^3$

73. $-4x^2y(-3x^2y^2) = 12x^{2+2}y^{1+2} = 12x^4y^3$

75. $(x^2)^4 = x^{2\bullet 4} = x^8$

77. $(m^{50})^{10} = m^{50\bullet 10} = m^{500}$

79. $(2a)^3 = 2^3a^3 = 8a^3$

81. $(xy)^4 = x^4y^4$

83. $(3s^2)^3 = 3^3s^{2\bullet 3} = 27s^6$

85. $(2s^2t^3)^2 = 2^2s^{2\bullet 2}t^{3\bullet 2} = 4s^4t^6$

87. $(x^2)^3(x^4)^2 = x^6x^8 = x^{14}$

89. $(c^5)^3(c^3)^5 = c^{15}c^{15} = c^{30}$

91. $(2a^4)^2(3a^3)^2 = 4a^8 9a^6 = 36a^{14}$

93. $(3a^3)^3(2a^2)^3 = 27a^9 \bullet 8a^6 = 216a^{15}$

95. $(x^2x^3)^{12} = (x^5)^{12} = x^{60}$

97. $(2b^4b)^5 = (2b^5)^5 = 32b^{25}$

Writing

99. Answers may vary.
101. Answers may vary.

Review

103. JEWELRY

$$\frac{18}{24} = \frac{3}{4}$$

105. $\dfrac{-25}{-5} = 5$

107. $2\left(\dfrac{12}{-3}\right) + 3(5) = -8 + 15 = 7$

109. $-x = -12$

$$\frac{-x}{-1} = \frac{-12}{-1}$$

$$x = 12$$

Section 6.4 Introduction to Polynomials

Vocabulary

1. A polynomial with one term is called a **monomial**.
3. A polynomial with two terms is called a **binomial**.

Concepts

5. This is a binomial.
7. This is a monomial.
9. This is a monomial.
11. This is a trinomial.
13. The degree is three.
15. The degree is two.
17. The degree is one.
19. The degree is seven.

Notation

21. $$\begin{aligned} 3a^2 + 2a - 7 &= 3(2)^2 + 2(2) - 7 \\ &= 3(4) + 4 - 7 \\ &= 12 + 4 - 7 \\ &= 16 - 7 \\ &= 9 \end{aligned}$$

Practice

23. $$\begin{aligned} 3x + 4 &= 3(3) + 4 \\ &= 9 + 4 \\ &= 13 \end{aligned}$$

25. $$\begin{aligned} 2x^2 + 4 &= 2(-1)^2 + 4 \\ &= 2(1) + 4 \\ &= 6 \end{aligned}$$

27. $$\begin{aligned} 0.5t^3 - 1 &= 0.5(4)^3 - 1 \\ &= 0.5(64) - 1 \\ &= 32 - 1 \\ &= 31 \end{aligned}$$

29. $\frac{2}{3}b^2 - b + 1 = \frac{2}{3}(3)^2 - 3 + 1$

$= \frac{2}{3}(9) - 3 + 1$

$= 6 - 3 + 1$

$= 4$

31. $-2s^2 - 2s + 1 = -2(-1)^2 - 2(-1) + 1$

$= -2(1) + 2 + 1$

$= -2 + 2 + 1$

$= 1$

33. $y = x^2$

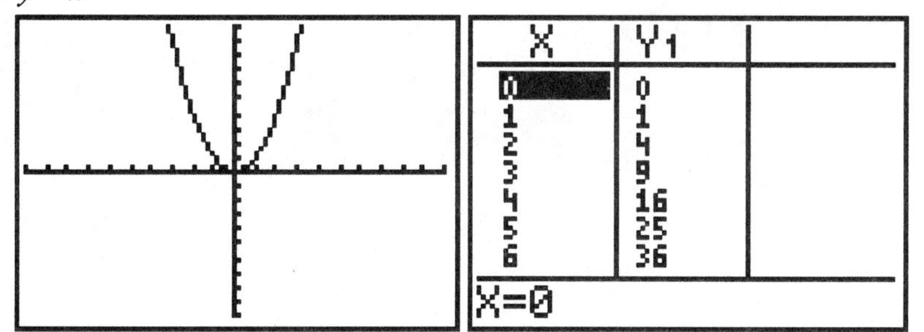

X	Y1
0	0
1	1
2	4
3	9
4	16
5	25
6	36

X=0

35. $y = \frac{1}{2}x^2$

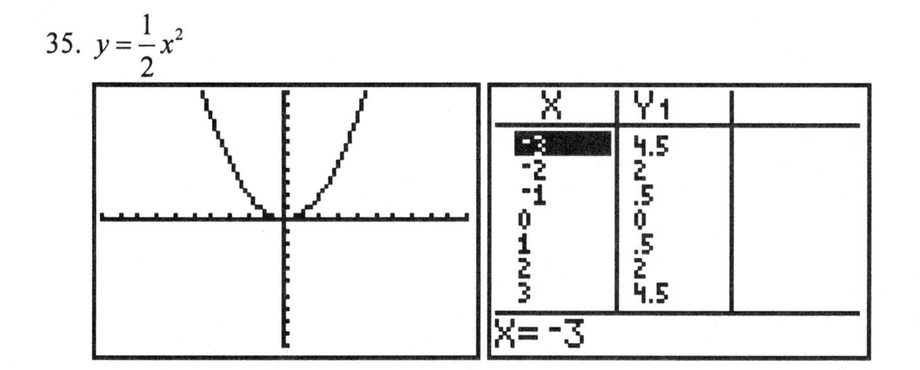

X	Y1
-3	4.5
-2	2
-1	.5
0	0
1	.5
2	2
3	4.5

X= -3

37. $y = -x^2 + 1$

X	Y1
-3	-8
-2	-3
-1	0
0	1
1	0
2	-3
3	-8

X= -3

39. $y = 2x^2 - 3$

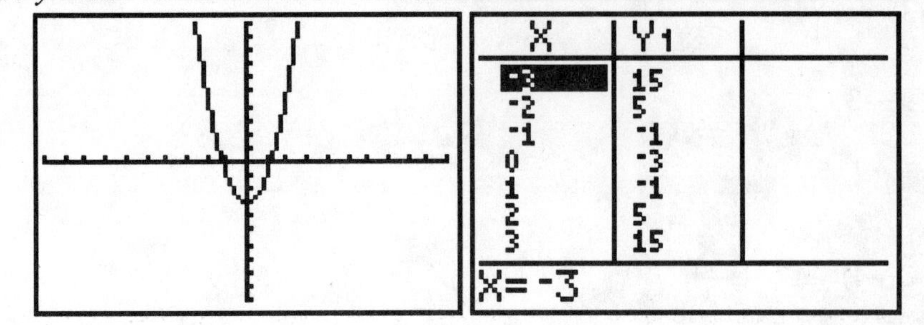

Applications

41. $h = -16t^2 + 64t$

$h = -16(0)^2 + 64(0)$

$h = 0$ ft

43. $h = -16t^2 + 64t$

$h = -16(2)^2 + 64(2)$

$h = -64 + 128$

$h = 64$ ft

45. $d = 0.04v^2 + 0.9v$

$d = 0.04(30)^2 + 0.9(30)$

$d = 36 + 27$

$d = 63$ ft

47. $d = 0.04v^2 + 0.9v$

$d = 0.04(60)^2 + 0.9(60)$

$d = 144 + 54$

$d = 198$ ft

49. SUSPENSION BRIDGE

x	0	2	4	-2	-4
y	0	1	4	1	4

Writing

51. Answers may vary.
53. Answers may vary.

Review

55. $\frac{2}{3}+\frac{4}{3}=\frac{6}{3}=2$

57. $\frac{36}{7}-\frac{23}{7}=\frac{13}{7}$

59. $\frac{5}{12}\bullet\frac{18}{5}=\frac{1}{6(2)}\bullet\frac{6(3)}{1}=\frac{3}{2}$

61. $x-4=12$

    $x=16$

63. $2(x-3)=6$

    $x-3=3$

    $x=6$

Section 6.5 Adding and Subtracting Polynomials

Vocabulary

1. If two algebraic terms have exactly the same variables and exponents, they are called **like** terms.

Concepts

3. To add two monomials, we add the **coefficients** and keep the same **variables** and exponents.
5. These are like terms, $3y+4y=7y$.
7. These are not like terms.
9. These are like terms, $3x^3+4x^3+6x^3=13x^3$.
11. These are like terms, $-5x^2+13x^2+7x^2=15x^2$.

Notation

13. $$\begin{aligned}(3x^2+2x-5)+(2x^2-7x)&=(3x^2+2x^2)+(2x-7x)+(-5)\\&=5x^2+(-5x)-5\\&=5x^2-5x-5\end{aligned}$$

Practice

15. $4y+5y=9y$

17. $-8t^2-4t^2=-12t^2$

19. $3s^2+4s^2+7s^2=14s^2$

21. $$\begin{aligned}(3x+7)+(4x-3)&=3x+4x+7-3\\&=7x+4\end{aligned}$$

23. $$\begin{aligned}(2x^2+3)+(5x^2-10)&=2x^2+5x^2+3-10\\&=7x^2-7\end{aligned}$$

25. $$\begin{aligned}(5x^3-4.2x)+(7x^3-10.7x)&=5x^3+7x^3-4.2x-10.7x\\&=12x^3-14.9x\end{aligned}$$

27. $$\begin{aligned}(3x^2+2x-4)+(5x^2-17)&=3x^2+5x^2+2x-4-17\\&=8x^2+2x-21\end{aligned}$$

29. $$\begin{aligned}(7y^2+5y)+(y^2-y-2)&=7y^2+y^2+5y-y-2\\&=8y^2+4y-2\end{aligned}$$

31. $(3x^2-3x-2)+(3x^2+4x-3)=3x^2+3x^2-3x+4x-3-2$

$=6x^2+x-5$

33. $(3n^2-5.8n+7)+(-n^2+5.8n-2)=3n^2-n^2-5.8n+5.8n+7-2$

$=2n^2+5$

35. $3x^2+4x+5$

$\underline{2x^2-3x+6}$

$5x^2+x+11$

37. $-3x^2-7$

$\underline{-4x^2-5x+6}$

$-7x^2-5x-1$

39. $-3x^2+4x+25.4$

$\underline{5x^2-3x-12.5}$

$2x^2+x+12.9$

41. $32u^3-16u^3=16u^3$

43. $18x^5-11x^5=7x^5$

45. $(4.5a+3.7)-(2.9a-4.3)=4.5a+3.7-2.9a+4.3$

$=1.6a+8$

47. $(-8x^2-4)-(11x^2+1)=-8x^2-4-11x^2-1$

$=-19x^2-5$

49. $(3x^2-2x-1)-(-4x^2+4)=3x^2-2x-1+4x^2-4$

$=7x^2-2x-5$

51. $(3.7y^2-5)-(2y^2-3.1y+4)=3.7y^2-5-2y^2+3.1y-4$

$=1.7y^2+3.1y-9$

53. $(2b^2+3b-5)-(2b^2-4b-9)=2b^2+3b-5-2b^2+4b+9$

$=7b+4$

55. $(5p^2-p+7.1)-(4p^2+p+7.1)=5p^2-p+7.1-4p^2-p-7.1$

$=p^2-2p$

57. $3x^2+4x-5$

$\underline{-(-2x^2-2x+3)}$

$5x^2+6x-8$

59. $-2x^2-4x+12$

$\underline{-(10x^2+9x-24)}$

$-12x^2-13x+36$

61. $4x^3-3x+10$

$\underline{-(5x^3-4x-4)}$

$-x^3+x+14$

Applications

63. VALUE OF A HOUSE

$y=700x+85,000$

$y=700(10)+85,000$

$y=\$92,000$

65. VALUE OF A HOUSE

$y=900x+102,000$

$y=900(12)+102,000$

$y=\$112,800$

67. VALUE OF TWO HOUSES

Method One – Use substitution into the equations

$y=700x+85,000$

$y=700(15)+85,000$

$y=\$95,500$

$y=900x+102,000$

$y=900(15)+102,000$

$y=\$115,500$

The value of the two houses is the sum of the above values, $95,500+115,500=\$211,000$.

Method Two – Use substitution into the equation of number 66.

$y=1,600(15)+187,000$

$y=\$211,000$

69. VALUE OF A CAR

$y=-800x+8,500$

71. VALUE OF TWO CARS

$$y = -800x + 8,500 - 1,100x + 10,200$$
$$= -1,900x + 18,700$$

Writing

73. Answers may vary.
75. Answers may vary.

Review

77. BASKETBALL SHOES

$14.6 - 13.8 = 0.8$ ounces

79. PANAMA CANAL

The ship must be lowered 54 feet by the Miraflores Lock system.

Section 6.6 Multiplying Polynomials

Vocabulary

1. A polynomial with one term is called a **monomial**.
3. A polynomial with **three** terms is called a trinomial.

Concepts

5. To multiply two monomials, multiply the **numerical** factors and then multiply the variable **factors**.
7. To multiply two binomials, multiply each **term** of one binomial by each term of the other binomial and combine **like** terms.

Notation

9. $3x(2x-5)=3x(2x)-3x(5)$
$=6x^2-15$

Practice

11. $3x^2(4x^3)=12x^5$

13. $3b^2(-2b)=-6b^3$

15. $-2x^2(3x^3)=-6x^5$

17. $\left(-\frac{2}{3}y^5\right)\left(\frac{3}{4}y^2\right)=-\frac{1}{2}y^7$

19. $3(x+4)=3x+12$

21. $-4(t+7)=-4t-28$

23. $3x(x-2)=3x^2-6x$

25. $-2x^2(3x^2-x)=-6x^4+2x^3$

27. $2x(3x^2+4x-7)=6x^3+8x^2-14x$

29. $-p(2p^2-3p+2)=-2p^3+3p^2-2p$

31. $3q^2(q^2-2q+7)=3q^4-6q^3+21q^2$

33. $(a+4)(a+5)=a^2+5a+4a+20=a^2+9a+20$

35. $(3x-2)(x+4)=3x^2+12x-2x-8=3x^2+10x-8$

37. $(2a+4)(3a-5)=6a^2-10a+12a-20=6a^2+2a-20$

39. $(2x+3)(2x+3)=4x^2+6x+6x+9=4x^2+12x+9$

41. $(2x-3)(2x-3)=4x^2-6x-6x+9=4x^2-12x+9$

43. $(5t+1)(5t+1)=25t^2+5t+5t+1=25t^2+10t+1$

45. $(2x+1)(3x^2-2x+1)=6x^3-4x^2+2x+3x^2-2x+1$
$=6x^3-x^2+1$

47. $(x-1)(x^2+x+1)=x^3+x^2+x-x^2-x-1$
$=x^3-1$

49. $(x+2)(x^2-3x+1)=x^3-3x^2+x+2x^2-6x+2$
$=x^3-x^2-5x+2$

51. $(4x+3)(x+2)=4x^2+8x+3x+6$
$=4x^2+11x+6$

53. $(4x-2)(3x+5)=12x^2+20x-6x-10$
$=12x^2+14x-10$

55. $(x^2-x+1)(x+1)=x^3+x^2-x^2-x+x+1$
$=x^3+1$

Applications

57. GEOMETRY
The area of the rectangle is $(x+2)(x-2)=x^2-2x+2x-4=(x^2-4)$ ft^2.

59. ECONOMICS

$$R=pq$$

$$r=\left(-\frac{x}{100}+30\right)x$$

$$r=-\frac{x^2}{100}+30x$$

Writing

61. Answers may vary.
63. Answers may vary.

Review

65. THE EARTH
 Four and ninety-one thousandths

67. $\frac{7}{64} = 0.109375$

69. $56.09 + 78 + 0.567 = 134.657$

71. $\sqrt{16} + \sqrt{36} = 4 + 6 = 10$

Chapter 6 Key Concept

1. The x-axis is the horizontal axis and the y-axis is the vertical axis.

3.

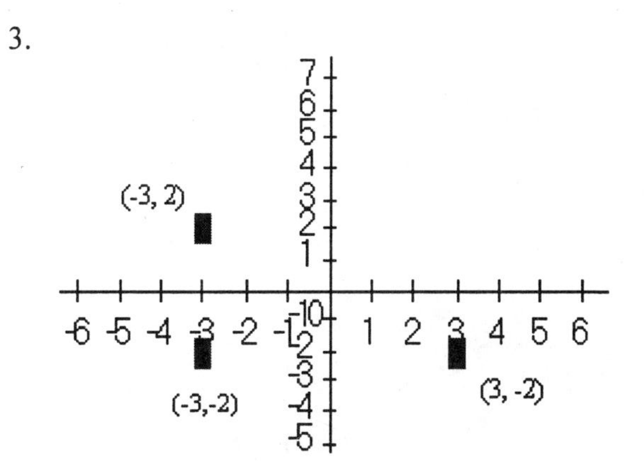

5. Quadrant II has negative x-coordinates and positive y-coordinates.

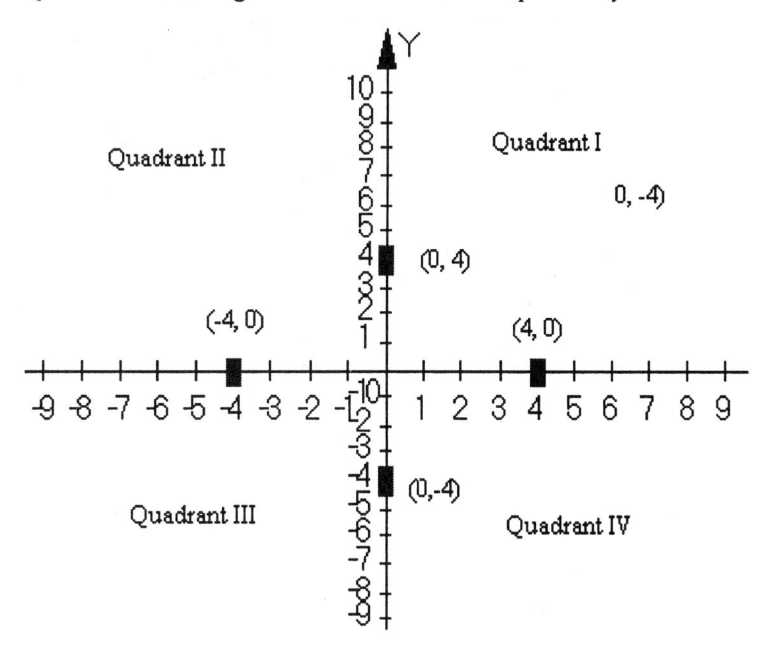

7. $2x - 4y = 8$

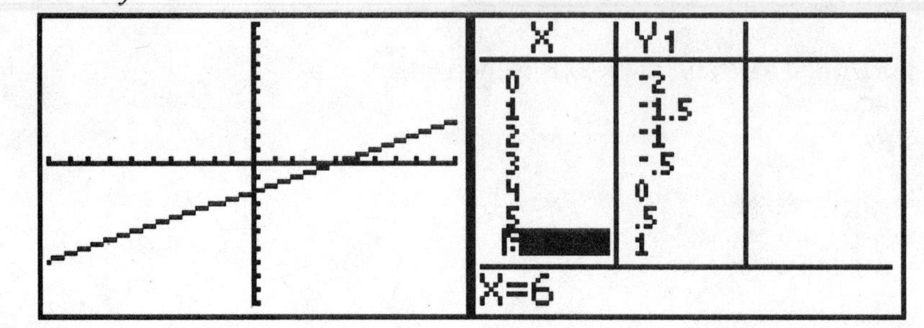

X	Y1
0	-2
1	-1.5
2	-1
3	-.5
4	0
5	.5
6	1

X=6

9. $y = x^2 + 1$

X	Y1
-3	10
-2	5
-1	2
0	1
1	2
2	5
3	10

X= -3

Chapter 6 Review

Section 6.1 The Rectangular Coordinate System

1. a. $2x+5y=-11$

   $2(2)+5(-3)=4-15=-11$

   They are a solution.

   b. $3x-5y=19$

   $3(-3)-5(2)=-9-10=-19\neq 19$

   They are not a solution.

3. a. 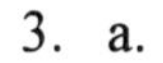

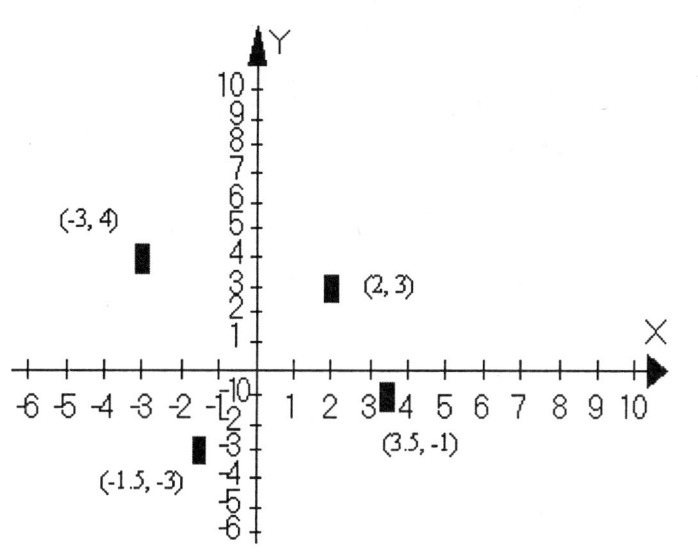

   b. $A(4,3), B(-3,3), C(-4,0), D(-1.5,-3.5), E(2.5,-1.5)$

5. Your seat is in the third row, second column.

Section 6.2 Graphing Linear Equations

7. a. $y = 2$

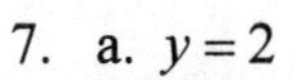

b. $x = 1$

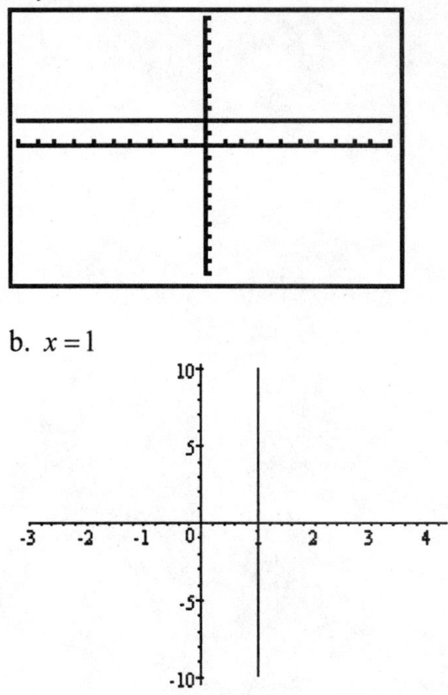

Section 6.3 Multiplication Rules for Exponents

9. a. $(4h)^3 = 4h \bullet 4h \bullet 4h$

   b. $5 \bullet 5 \bullet d \bullet d \bullet d \bullet m \bullet m \bullet m \bullet m = 5^2 d^3 m^4$

11. a. $2b^2 4b^5 = 8b^7$

   b. $-6x^3(4x) = -24x^4$

   c. $-2f^2(-4f)(3f^4) = 24f^7$

   d. $-ab \bullet b \bullet a = -a^2b^2$

   e. $xy^4 \bullet xy^2 = x^2y^6$

   f. $(mn)(mn) = m^2n^2$

   g. $3z^3 \bullet 9m^3z^4 = 27m^3z^7$

   h. $-5cd(4c^2d^5) = -20c^3d^6$

13. a. $(c^4)^5(c^2)^3 = c^{20}c^6 = c^{26}$

   b. $(3s^2)^3(2s^3)^2 = 27s^6 \bullet 4s^6 = 108s^{12}$

   c. $(c^4c^3)^2 = (c^7)^2 = c^{14}$

   d. $(2xx^2)^3 = (2x^3)^3 = 8x^9$

Section 6.4 Introduction to Polynomials

15. a. This has degree three.
    b. This has degree four.
    c. This has degree five.

17. a. $y = x^2 - 3$

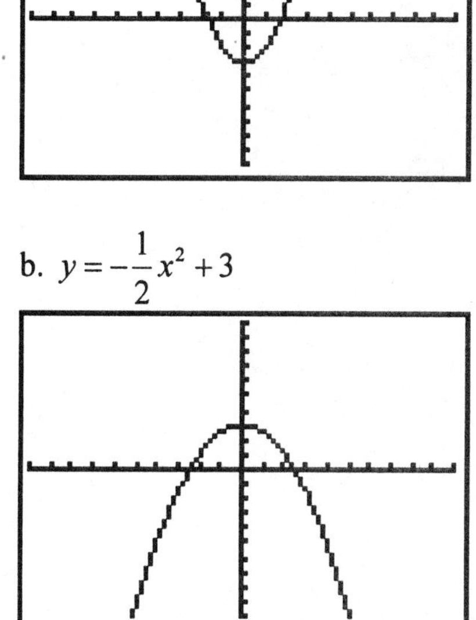

    b. $y = -\frac{1}{2}x^2 + 3$

Section 6.5 Adding and Subtracting Polynomials

19. a. $(3x - 1) + (6x + 5) = 9x + 4$
    b. $(3x^2 - 2x + 4) + (-x^2 - 1) = 2x^2 - 2x + 3$

21. a. $16p^3 - 9p^3 = 7p^3$
    b. $4y^2 - 9y^2 = -5y^2$

23. a. $(5x - 2) - (3x + 5) = 5x - 2 - 3x - 5 = 2x - 7$
    b. $(3x^2 - 2x + 7) - (-5x^2 + 3x - 5) = 3x^2 - 2x + 7 + 5x^2 - 3x + 5$
    $= 8x^2 - 5x + 12$

Section 6.6 Multiplying Polynomials

25. a. $2x^2(3x+2)=6x^3+4x^2$

    b. $-5t^3(7t^2-6t-2)=-35t^5+30t^4+10t^3$

27. a. $(5x-2)(3x+5)=15x^2+25x-6x-10$
    $=15x^2+19x-10$

    b. $(3x+2)(5x-5)=15x^2-15x+10x-10$
    $=15x^2-5x-10$

29. a. $(5x^2-2x+3)(3x+5)=15x^3+25x^2-6x^2-10x+9x+15$
    $=15x^3+19x^2-x+15$

    b. $(3x^2-2x-1)(5x-2)=15x^3-6x^2-10x^2+4x-5x+2$
    $=15x^3-16x^2-x+2$

Chapter 6 Test

1. $4x+5y=6$
   $4(-1)+5(2)=-4+10=6$

3. $(0,-2)$ since $0-2y=4 \rightarrow y=-2$
   $(4,0)$ since $x-2(0)=4 \rightarrow x=4$
   $(2,-1)$ since $2-2y=4 \rightarrow y=-1$

5. PANTS SALE
   $(30,32),(30,34),(31,34),(38,30)$

7. $A(0,0), B(2.5,3.5), C(-3,-2), D(0,-2)$

9. a. The x-intercept is (2, 0).
   b. The y-intercept is (0, -4).

11. $y=-2$

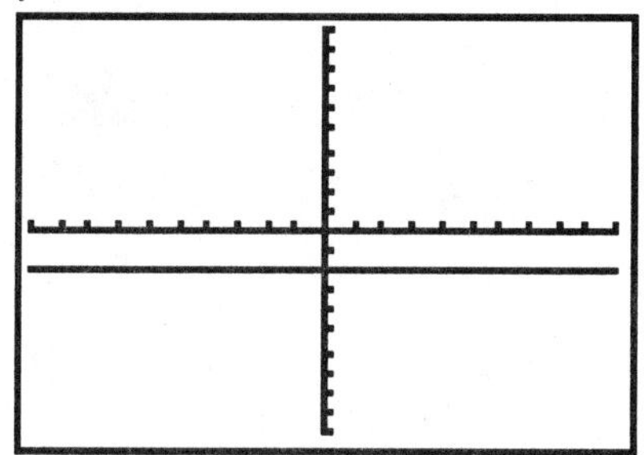

13. a. $h^2h^4=h^6$
    b. $-7x^3(4x^2)=-28x^5$
    c. $b^2bb^5=b^8$
    d. $-3g^2k^3(-8g^3k^{10})=24g^5k^{13}$

15. This is a binomial.

17. This has degree six.

19. $3x^2-2x+4$
    $3(3)^2-2(3)+4=27-6+4=25$

21. $y = 2x^2$

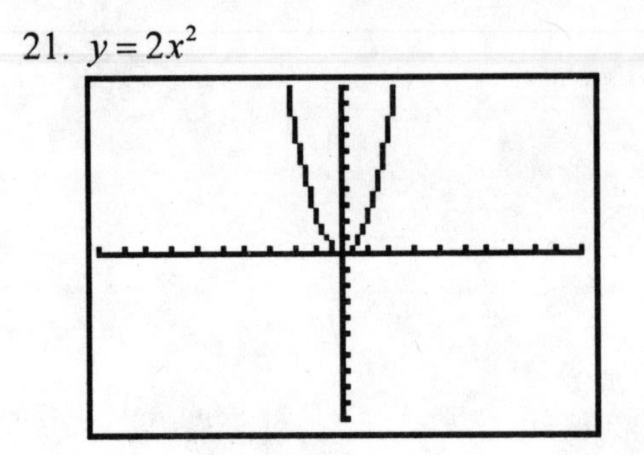

23. $(3x^2 + 2x) + (2x^2 - 5x + 4) = 5x^2 - 3x + 4$

25. $(2.1p^2 - 2p - 2) - (3.3p^2 - 5p - 2) = 2.1p^2 - 2p - 2 - 3.3p^2 + 5p + 2$
$= -1.2p^2 + 3p$

27. $(-2x^3)(4x^2) = -8x^5$

29. $(2x - 5)(3x + 4) = 6x^2 + 8x - 15x - 20$
$= 6x^2 - 7x - 20$

31. These are not the same points as (1, -2) lies in quadrant IV and (-2, 1) lies in quadrant II.

Chapters 1 - 6 Cumulative Review

1. 6,246,000
3. This rectangle has perimeter $P = 2(8) + 2(3) = 16 + 6 = 22$ meters.

5. PARKING
   $Area = length \bullet width$

Type	Length (ft)	Width (ft)	Area (ft^2)
Standard	20	9	180
Standard to a wall	20	10	200
Parallel	25	10	250
Compact	17	8	136

7. $120 = 2^3 \bullet 3 \bullet 5$

9. LAKE TAHOE
   Since the 1960's the visibility is $105 - 66 = 39$ feet less.

11. $$\begin{aligned} 12 - 2[1 - (-8 + 2)] &= 12 - 2[1 - (-6)] \\ &= 12 - 2[7] \\ &= 12 - 14 \\ &= -2 \end{aligned}$$

13. $5x - 11x = -6x$

15. $$\begin{aligned} 4x + 3 &= 11 \\ 4x &= 8 \\ x &= 2 \end{aligned}$$

17. $$\begin{aligned} \frac{t}{3} + 2 &= -4 \\ \frac{t}{3} &= -6 \\ t &= -18 \end{aligned}$$

19. $\frac{5}{10b^3} \bullet 2b^2 = \frac{10b^2}{10b^3} = \frac{1}{b}$

21. $34\frac{1}{9} - 13\frac{5}{6} = 34\frac{2}{18} - 13\frac{15}{18}$

$= 33\frac{20}{18} - 13\frac{15}{18}$

$= 20\frac{5}{18}$

23. $\frac{7}{8}t = -28$

$t = \frac{8}{7}(-28)$

$t = -32$

25. PAPER SHREDDER

The shredder will create $\frac{8\frac{1}{2}}{\frac{1}{4}} = 8\frac{1}{2} \bullet 4 = 34$ strips.

27. 57.57

29. $287.23 - 179.97 = 107.26$

31. $3.8\overline{)17.746}$ with quotient 4.67

33. $5\frac{5}{8} \approx 5.6$

35. $3.2x = 74.46 - 1.9x$

$5.1x = 74.46$

$x = 14.6$

37. $-2(x - 2.1) = -2.4$

$x - 2.1 = 1.2$

$x = 3.3$

39. EARTHQUAKES

The mean of this data is 8.2, the median is 8.0, and the mode is 7.9.

9.2	
8.8	
8.7	
8.3	
8.3	
8.2	
8.2	
8	
7.9	
7.9	
7.9	
7.9	
7.9	
7.8	
7.8	
8.186667	Mean
8	Median
7.9	Mode

41. PETITION DRIVE

Creating an equation, $P = 20 + 0.05s$

$$60 = 20 + 0.05s$$
$$40 = 0.05s$$
$$800 = s$$

She must collect 800 signatures.

43. $\sqrt{121} = 11$

45. $\sqrt{0.25} = 0.5$

47. $4x - 5y = -23$

$4(-2) - 5(3) = -8 - 15 = -23$

49. $p^4 p^5 = p^9$

51. $(p^3q^2)(p^3q^4) = p^6q^6$

53. $(3x^2 - 5x) - (2x^2 + x - 3) = 3x^2 - 5x - 2x^2 - x + 3$

$= x^2 - 6x + 3$

Section 7.1 Percents, Decimals, and Fractions

Vocabulary

1. **Percent** means part per one hundred.

Concepts

3. To write a percent as a fraction, drop the percent symbol and write the given number over **100**.
5. To change a decimal to a percent, multiply the decimal by 100 by moving the decimal point two places to the **right**, and then insert a % symbol.
7. a. 0.84, 84%, $\frac{84}{100}=\frac{21}{25}$

   b. 16% of the figure is not shaded.

Practice

9. $17\% = \frac{17}{100}$

11. $5\% = \frac{5}{100} = \frac{1}{20}$

13. $60\% = \frac{60}{100} = \frac{3}{5}$

15. $125\% = \frac{125}{100} = \frac{5}{4}$

17. $\frac{2}{3}\% = \frac{\frac{2}{3}}{100} = \frac{2}{300} = \frac{1}{150}$

19. $5\frac{1}{4}\% = \frac{\frac{21}{4}}{100} = \frac{21}{400}$

21. $0.6\% = \frac{\frac{6}{10}}{100} = \frac{3}{500}$

23. $1.9\% = \frac{1\frac{9}{10}}{100} = \frac{\frac{19}{10}}{100} = \frac{19}{1{,}000}$

25. $19\% = 0.19$

27. $6\% = 0.06$

29. $40.8\% = 0.408$

31. $250\% = 2.50$

33. $0.79\% = 0.0079$

35. $\frac{1}{4}\% = 0.25\% = 0.0025$

37. $0.93 = 93\%$

39. $0.612 = 61.2\%$

41. $0.0314 = 3.14\%$

43. $8.43 = 843\%$

45. $50 = 5{,}000\%$

47. $9.1 = 910\%$

49. $\frac{17}{100} = 17\%$

51. $\frac{4}{25} = \frac{16}{100} = 16\%$

53. $\frac{2}{5} = \frac{40}{100} = 40\%$

55. $\frac{21}{20} = \frac{105}{100} = 105\%$

57. $\frac{5}{8} = 62.5\%$

59. $\frac{3}{16} = 18.75\%$

61. $\frac{2}{3} = 66\frac{2}{3}\%$

63. $\frac{5}{6} = 83\frac{1}{3}\%$

65. $\frac{1}{9} \approx 11.11\%$

67. $\frac{5}{9} \approx 55.56\%$

Applications

69. U.N. SECURITY COUNCIL

a. $\frac{15}{188}$ belong to the Security Council

b. $\frac{15}{188} \approx 8\%$

71. PIANO KEYS

a. $\frac{36}{88} = \frac{9}{22}$

b. $\frac{9}{22} \approx 41\%$

73. a. There are 29 bones in the spinal column, $\frac{5}{29}$ are lumbar.

b. $\frac{5}{29} \approx 17\%$

c. $\frac{7}{29} \approx 24\%$

75. STEEP GRADE

Over 100 feet, a 5% grade rises 5 feet since $\frac{5}{100} = 5\%$..

77. IVORY SOAP

$99\frac{44}{100}\% = 99.44\% = 0.9944$

79. BASKETBALL STANDINGS

The winning percentage is presented as a decimal. It is equal to 89.6%.

81. SKIN

The torso is 27.5 % since the total percents of the entire body should sum to 100%.

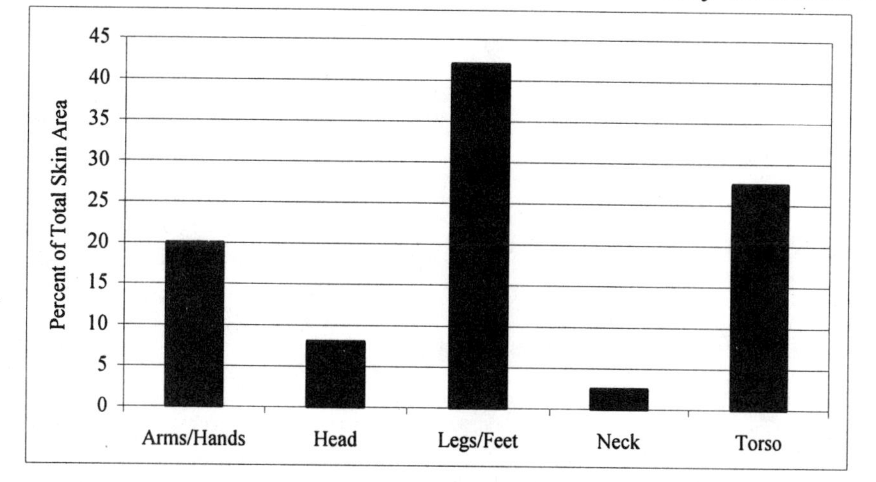

83. CHARITY

$\frac{92}{100} = 92\%$ of the money spent by the Red Cross went to programs and services.

85. BIRTHDAY

$\frac{1}{365} \approx 0.27\%$

Writing

87. Answers may vary
89. Answers may vary
91. Answers may vary

Review

93. $-\frac{2}{3}x = -6$

$$x = \frac{18}{2}$$

$$x = 9$$

95. $y = 2x + 3; (2,7), (4,11), (0,3)$

$y = 2(2) + 3 = 7 \rightarrow (2,7)$

$y = 2(4) + 3 = 11 \rightarrow (4,11)$

$y = 2(0) + 3 = 3 \rightarrow (0,3)$

97. $(x+1)(x+2)$

$=x^2+2x+x+2$

$=x^2+3x+2$

Section 7.2 Solving Percent Problems

Vocabulary

1. $n = 0.10(50)$
3. $48 = n \bullet 47$
5. In a circle **graph**, pie-shaped wedges are used to show the division of a whole quantity into its component parts.

Concepts

7. Change to a decimal.
   a. $12\% = 0.12$
   b. $5.6\% = 0.056$
   c. $125\% = 1.25$
   d. $\frac{1}{4}\% = 0.25\% = 0.0025$

9. 120% of 55 is more than 55

11. a. 100% of 25 is 25
   b. 132 of 132 is 100%
   c. 87 is 87% of 100

13. VIDEO GAMES
   Nintendo 64 has $100\% - 4\% - 63\% = 33\%$ of the market.

Notation

15. Translate each.
   a. "of" translates to multiply
   b. "is" translates to equal
   c. "what number" translates to a variable, often denoted with x

Practice

17. $x = 0.36(250)$

   $x = 90$

19. $16 = x(20)$

   $0.08 = x$

   $x = 80\%$

21. $7.8 = 0.12x$

   $65 = x$

23. $x = 0.008(12)$

$x = 0.096$

25. $0.5 = x(40,000)$

$0.0000125 = x$

$x = 0.00125\%$

27. $3.3 = 0.075x$

$44 = x$

29. $x = 0.0725(600)$

$x = 43.5$

31. $1.02(105) = x$

$107.1 = x$

33. $0.\bar{3}x = 33$

$x = 99$

35. $0.095x = 5.7$

$x = 60$

37. $x(8,000) = 2,500$

$x = 0.3125$

$x = 31.25\%$

39.

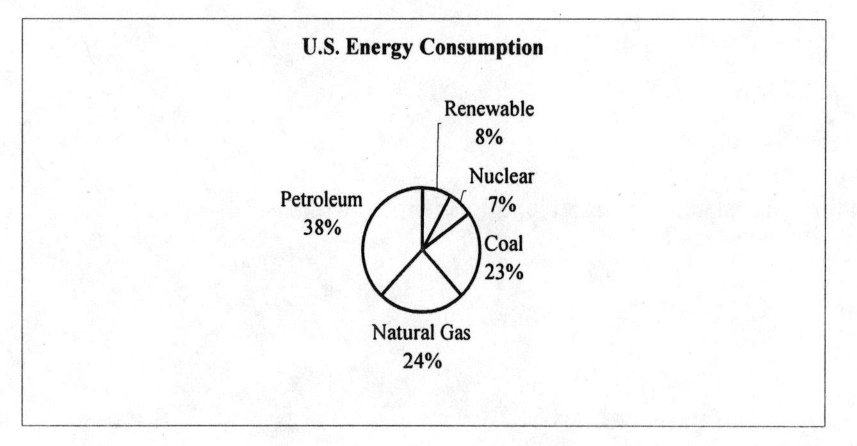

Applications

41. CHILD CARE

$0.70c = 84$

$c = 120$

The daycare center could enroll 120 youngsters.

43. GOVERNMENT SPENDING

$0.37(1{,}650 \text{ billion}) = 610.5 \text{ billion}$

$610.5 billion was spent on Social Security, Medicare and other retirement programs

45. THE INTERNET

$0.24(50{,}000) = 12{,}000 = 12\text{K}$

$50\text{K} - 12\text{K} = 38\text{K}$

There are 38K bytes left to download.

47. PRODUCTION PROMOTION

$0.25s = 6$

$s = 24$

There are 24 ounces in the large bottle.

49. DRIVER'S LICENSE

$0.7(40) = 28$

To pass the test 28 questions must be answered correctly, so he passed.

51. MIXTURES

Gallons Solution	% Acid	Gallons Acid
60	50%	$0.5(60) = 30$
40	30%	$0.3(40) = 12$

53. MAKING COPIES

$1.80(1.5) = 2.7$

The height of the print on the copy will be 2.7 inches.

55. INSURANCE

$200 is what percentage of $4,000

$200 = d(4{,}000)$

$0.05 = d$ or 5%

57. A MAJORITY

The total votes cast were $8,501+3,614+2,630+2,432=17,177$.

50% of 17,177 is 8,589, so there must be a runoff election.

Writing

59. Answers may vary
61. Answers may vary

Review

63. $2.78+6+9.09+0.3=18.17$

65. 5.001 is closer to 5 on a number line since $5.001-5=0.001$ and $5-4.9=0.10$.

67. $34.5464 \cdot 1,000=34,546.4$

69. $0.4x+1.2=-7.8$

$$0.4x=-9$$

$$x=-22.5$$

Section 7.3 Applications of Percents

Vocabulary

1. Some salespeople are paid on **commission**. It is based on a percent of the total dollar amount of the goods or services they sell.
3. The difference between the original price and the sale price of an item is called the **discount**.

Concepts

5. If there has been a 100% increase then the membership has doubled.

Applications

7. STATE SALES TAX
   The sales tax is $0.0475(900) = \$42.75$.

9. ROOM TAX
   \$10.32 is what percent of \$129
   $10.32 = t(129)$
   $0.08 = t$ an 8% room tax

11. SALES RECEIPT
    Subtotal $8.97 + 9.87 + 28.50 = \$47.34$
    Sales Tax $0.06(47.34) = \$2.84$
    Total $\$47.34 + \$2.84 = \$50.18$

13. SALES TAX HIKE
    1% of \$15,000 is $0.01(15{,}000) = \$150$

15. PAYCHECK
    Federal Tax, \$28.80 is what percent of \$360
    $28.80 = t(360)$
    $0.08 = t$ or 8% tax

    Medicare Taxes, \$5.04 is what percent of \$360
    $5.04 = m(360)$
    $0.014 = m$ or 1.4% tax

    Worker's Compensation, \$4.32 is what percent of \$360
    $4.32 = w(360)$
    $0.012 = w$ or 1.2% tax

17. OVERTIME

480 less 25% of 480 is $480 - 0.25(480) = 360$.

The company target is 360 hours of overtime next month.

19. REDUCED CALORIES

150 calories less 36% of 150 calories is $150 - 0.36(150) = 96$.

There are 96 calories in the new product.

21. ENDANGERED SPECIES

There was a decline during 1995-1996.

$1{,}599 - 1{,}523 = 76$

$76 = x(1{,}599)$, x represents the percent decrease

$0.04753 \approx x$

During this time frame the decrease was approximately 5%.

23. CAR INSURANCE

The difference in the premium was $40.

$40 is what percent of $400

$40 = c(400)$

$0.10 = c$, a 10% discount

25. LAKE SHORELINE

The shoreline increased by 1.8 miles.

1.8 miles is what percent of 5.8 miles

$1.8 = s(5.8)$

$0.31034 \approx s$

There was a 31% increase.

27. EARTH MOVING

a. The soil increases from 1 yd^3 to 1.25 yd^3, a difference of 0.25 yd^3.

0.25 yd^3 is what percent of 1 yd^3

$0.25 = x(1)$

$0.25 = d$ or a 25% increase

b. The soil decreases from 1.25 yd^3 to 0.80 yd^3, a difference of 0.45 yd^3.

0.45 yd^3 is what percent of 1.25 yd^3

$0.45 = d(1.25)$

$0.36 = d$ or a 36% decrease

29. REAL ESTATE

6% of $98,500 is $0.06(98{,}500) = \$5{,}910$.

Each agent received half of this amount or $2,955.

31. SPORTS AGENT

$37,500 is what percent of $2,500,000

$37{,}500 = f(2{,}500{,}000)$

$0.015 = f$ or a rate of 1.5%

33. CONCERT PARKING

The total parking intake is $\$6(6000) = \$36{,}000$.

$0.\bar{3}(36{,}000) = \$12{,}000$

The vendor can expect $12,000.

35. WATCH SALE

The regular price is $\$29.95 + \$10 = \$39.95$.

The discount is found considering $10 is what percent of $39.95

$10 = w(39.95)$

$0.25 \approx w$ or a 25% discount

37. RING SALE

80% of what is $149.99

$0.8r = 149.99$

$r \approx 187.49$

The original price was $187.49.

39. VCR SALE

The sale price is $\$399.97 - \$50 = \$349.97$.

The percent decrease is found considering $50 is what percent of $399.97.

$50 = v(399.97)$

$0.125 \approx v$ or a 12.5% discount

41. REBATE

The discount is $3.60.

The new price is $\$15.48 - \$3.60 = \$11.88$.

The percent decrease is found considering $3.60 is what percent of $ 15.48.

$3.60 = x(15.48)$

$0.2326 \approx x$ or a 23% decrease

43. TV SHOPPING

The ring sells for 45% of $170 or $0.45(170) = \$76.50$.

Writing

45. Answers may vary

47. Answers may vary

Review

49. $-5(-5)(-2) = -50$

51. $2(-2)^2(-2-1) = 8(-3) = -24$

53. There are $12d$ eggs is d dozens.

55. $|-5-8| = |-13| = 13$

57. $A(-3,4), B(4,3.5), C(-2,-\frac{5}{2}), D(0,-4), E(\frac{3}{2},0), F(3,-4)$

A ▫ 4 ▫B
2
E
-4 -2 0 2 4
-2
C ▫
-4 D ▫ F

Estimation
An estimated solution is given as well as the exact solution for comparison.

Study Set

1. COLLEGE COURSES
   $0.2(800) = 160$; 20% of 815 students is 163 students enrolled in science

3. DISCOUNT
   $0.3(200) = 60$; 30% of $196.88 is $59.06

5. FIRE DAMAGE
   $0.5(108,000) = 54,000$; 50% of $107,809 is $53,904.50

7. WEIGHTLIFTING
   $2(160) = 320$; 200% of 158 is 316

9. TRAFFIC STUDY
   20% of 650 is 130 since 10% is 65, double that for 20%

11. NO-SHOWS
    $0.30(68) = 20.4$ or 21 people; 31% of 68 is 21.08 or 22 people

13. INTERNET SURVEY
    $0.60(30,000) = 18,000$; 58% of 28,657 is 16,621.06 or 16,622 people

15. VOTING
    50% of 6,200 is 3,100; 48% of 6,200 is 2,976

Section 7.4 Interest

Vocabulary

1. In banking the original amount of money borrowed or deposited is known as the **principal**.
3. The percent that is used to calculate the amount of interest to be paid is called the **interest** rate.
5. Interest computed only on the original principal is called **simple** interest.

Concepts

7. a. $7\% = 0.07$
   b. $9.8\% = 0.098$
   c. $6\frac{1}{4}\% = 0.0625$

9. $I = Prt$
   $I = (10,000)(0.06)(3)$
   $I = \$1,800$

11. a. Compound interest is illustrated.
    b. The original principal was $1,000.
    c. Interest was calculated four times.
    d. The first compounding earned $50.
    e. The money was invested for one year.

Notation

13. Multiplication is implied by the *Prt.*

Applications

15. RETIREMENT INCOME
    At the end of year one there will be the original principal plus the interest,
    $5,000 + 0.06(5,000) = \$5,300$

17. REMODELING
    In two years the homeowner will pay $8,000(0.092)(2) = \$1,472$.

19. MEETING A PAYROLL

The business had to repay

$$P + I = 4,200 + 4,200(0.18)\left(\frac{30}{365}\right)$$
$$= \$4,200 + \$62.14$$
$$= \$4,262.14$$

21. SAVINGS ACCOUNT

P	r	t	$I = Prt$
\$10,000	0.0725	2	$10,000(0.0725)(2) = \$1,450$

23. LOAN APPLICATION

Amount \$1,200
Length 2 years
Rate 8%
Interest $1,200(0.08)(2) = \$192$
Repayment $1,200 + 192 = \$1,392$
Payments monthly

Borrower agrees to pay 24 equal payments of $\frac{1,392}{24} = \$58$ to repay the loan.

25. LOW-INTEREST LOAN

The total repayment is

$$P + I = 18,000,000 + 18,000,000(0.023)(2)$$
$$= 18,000,000 + 828,000$$
$$= \$18,828,000$$

27. COMPOUNDING ANNUALLY

Using $A = P\left(1 + \frac{r}{n}\right)^{nt}$, the account balance will be $A = 600\left(1 + \frac{0.08}{1}\right)^{1(3)}$

$A = \$755.83$

29. COLLEGE FUND

Using $A = P\left(1 + \frac{r}{n}\right)^{nt}$, the account balance will be $A = 1,000\left(1 + \frac{0.06}{365}\right)^{365(4)}$

$A = \$1,271.22$

31. TAX REFUND

Using $A = P\left(1+\frac{r}{n}\right)^{nt}$, the account balance will be $A = 545\left(1+\frac{0.046}{365}\right)^{365(1)}$.

$A = \$570.65$

33. LOTTERY

Using $A = P\left(1+\frac{r}{n}\right)^{nt}$, the account balance will $A = 500{,}000\left(1+\frac{0.06}{365}\right)^{365(1)}$.

$A = \$530{,}915.66$

The amount earned will be $\$530{,}915.66 - \$500{,}000 = \$30{,}915.66$.

Writing

35. Answers may vary
37. Answers may vary

Review

39. $\sqrt{\frac{1}{4}} = \frac{1}{2}$

41. $y = 2x - 10$ using the point $(2,-3)$

$-3 \neq 2(2) - 10$

This point is not on the line.

43. $\frac{2}{3}x = -2$

$x = \frac{-6}{2}$

$x = -3$

45. This polynomial has three terms.

47. This point lies in quadrant three.

Chapter 7 Key Concepts

1. $\frac{10}{24}=\frac{5}{12}$; simplifying fractions

3. $2x+3x=5x$; combining like terms

5. $x^3 \bullet x^2 = x^5$; like bases so add exponents

7. $\frac{2}{3}=66\frac{2}{3}\%$; change the fraction to a decimal, move the decimal two places to the right and insert the % sign.

9. $4x^2+1-2x^2=2x^2+1$; combine like terms

11. $\frac{6}{6}=1$; a number divided by itself is one, except 0

13. $(-5)(-6)=30$; product of like sign integers is positive

15. $\frac{2x}{2}=x$; the quotient of 2 and 2 is one

17. $2+3 \bullet 5=2+15=17$; multiplication then addition

19. $2(x+5)=2x+10$; distributive property of multiplication over addition

21. $\frac{2}{3} \bullet \frac{3}{2}=1$; product of reciprocals is one

Chapter 7 Review

Section 7.1 Percents, Decimals, and Fractions

1. a. $39\%, 0.39, \frac{39}{100}$

   b. $111\%, 1.11, \frac{111}{100}$

3. a. $15\% = \frac{15}{100} = \frac{3}{20}$

   b. $120\% = \frac{120}{100} = \frac{6}{5}$

   c. $9\frac{1}{4}\% = \frac{37}{4}\% = \frac{\frac{37}{4}}{100} = \frac{37}{400}$

   d. $0.1\% = \frac{\frac{1}{10}}{100} = \frac{1}{1,000}$

5. a. $0.83 = 83\%$
   b. $0.625 = 6.25\%$
   c. $0.051 = 5.1\%$
   d. $6 = 600\%$

7. a. $\frac{1}{3} = 33\frac{1}{3}\%$

   b. $\frac{5}{6} = 0.8\overline{3} = 83\frac{1}{3}\%$

9. BILL OF RIGHTS
   17 of 27 were adopted after the Bill of Rights
   $\frac{17}{27} = 0.\overline{629} \approx 63\%$

Section 7.2 Solving Percent Problems

11. The amount is 15, the base is 45 and the percent is $33\frac{1}{3}$.

13. a. 200 is 40% of 500; $n = 0.40(500) = 200$
    b. 16% of 125 is 20; $(0.16)x = 20$
    c. 1.4 is 1.75% of 80; $1.4 = x(80)$
    d. $66\frac{2}{3}$% of 3,150 is 2,100; $x = 0.6\bar{6}(3,150) = 2,100$
    e. 220% of 55 is 121; $x = 2.20(55) = 121$
    f. 0.05% of 60,000 is 30; $x = 0.0005(60,000) = 30$

15. HOME SALES
    $0.75h = 51$
    $h = 68$
    There were 68 homes in this subdivision.

17. TIPPING
    15% of \$36.20 is $0.15(36.30) = \$5.43$.

19. EARTH'S SURFACE
    70.9% of 196,800,000 is $0.709(196,800,000) = 139,531,200$ mi^2

Section 7.3 Applications of Percent

21. SALES TAX RATE
    $12,300t = 492$
    $t = 0.04$
    The tax rate is 4%.

23. TROOP SIZE
    The increase was $12,500 - 10,000 = 2,500$ troops.
    $2,500 = x(10,000)$
    $x = 0.25$ or a 25% increase

25. TOOL CHEST
    The discount was \$50 so the original price was $\$139.99 + \$50 = \$189.99$.
    The discount rate was found answering \$50 is what percent of \$189.99.
    $50 = 189.99t$
    $0.263172 \approx t$ or approximately a 26% discount

Section 7.4 Interest

27. CODE VIOLATIONS

The interest was $I = 10{,}000(0.125)\left(\frac{90}{365}\right) \approx \$308.22$.

The total repayment was $\$10{,}000 + \$308.22 = \$10{,}308.22$.

29. Using $A = P(1+\frac{r}{n})^{nt}$, $A = 2{,}000\left(1+\frac{0.07}{2}\right)^{2(1)}$

$A = \$2{,}142.45$

31. CASH GRANT

Using $A = P(1+\frac{r}{n})^{nt}$, $A = 500{,}000\left(1+\frac{0.083}{365}\right)^{365(1)}$.

$A \approx \$543{,}265.78$

The interest earned is $\$543{,}265.78 - \$500{,}000 = \$43{,}265.78$.

Chapter 7 Test

1. 61% of the figure is shaded, 0.61, $\frac{61}{100}$

3. a. $67\% = 0.67$
   b. $12.3\% = 0.123$
   c. $9\frac{3}{4}\% = 0.0975$

5. a. $0.19 = 19\%$
   b. $3.47 = 347\ \%$
   c. $0.005 = 0.5\%$

7. $\frac{7}{30} = 0.2\bar{3} = 23.33\%$

9. $\frac{2}{3} = 0.\bar{6} = 66\frac{2}{3}\%$

11. SHRINKAGE
    a. 3% of 34 is $0.03(34) = 1.02$ inches
    b. $34 - 1.02 = 32.98$ inches

13. TIPPING
    $0.15(25.40) = \$3.81$ tip

15. SWIMMING WORKOUT
    20% of the total number of laps is 18
    $0.2L = 18$
    $L = 90$ laps

17. $x = 0.24(600)$
    $x = 144$

19. HOMEOWNER'S INSURANCE
    4% of \$898 is $0.04(898) = \$35.92$

21. CAR WAX SALE
    The sale price is $\$14.95 - \$3 = \$11.95$.
    The discount is \$3.
    The discount rate is $3 = 14.95w$
    $0.2007 \approx w$, approximately 20%

23. $I = 3,000(0.05)(1) = \$150$

25. POLITICAL AD
37% of what is not clear in this advertisement.

Chapters 1 – 7 Cumulative Review Exercises

1. SHAQUILLE

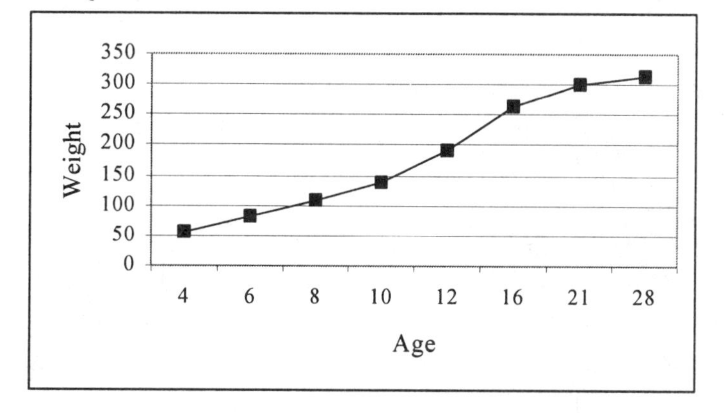

3. a. The factors of 40 are {1, 2, 4, 5, 8, 10, 20, 40}

   b. The prime factorization of 40 is $40 = 2^3 \cdot 5$

5. PAINTING

   An 8-foot square tarp will cover an area of $8 \cdot 8 = 64$ ft^2.

7. $12 - 2[-8 - 2^4(-1)]$
   $= 12 - 2[-8 - 16(-1)]$
   $= 12 - 2[-8 + 16]$
   $= 12 - 2[8]$
   $= 12 - 16$
   $= -4$

9. $6 = 2 - 2x$
   $4 = -2x$
   $-2 = x$

11. FRUIT STORAGE

$$C = \frac{5(59-32)}{9}$$

$$C = 15^0$$

13. SPELLING

$\frac{4}{11}$ of the letters in Mississippi are vowels

15. $-\frac{16a}{35}\bullet\frac{25}{48a^2}$

$=-\frac{16a}{7\bullet5}\bullet\frac{5\bullet5}{16\bullet3a^2}$

$=-\frac{1}{7}\bullet\frac{5}{3a}$

$=-\frac{5}{21a}$

17. $\frac{4}{m}+\frac{2}{7}=\frac{28}{7m}+\frac{2m}{7m}=\frac{28+2m}{7m}$

19. $\frac{5}{6}y=-25$

$y=-25\left(\frac{6}{5}\right)$

$y=-30$

21. $78.1-7.81=70.29$

23. $0.752(1{,}000)=752$

25. $\frac{3.6-(-1.5)}{0.5(-1.5)-0.4(3.6)}=\frac{5.1}{-2.19}\approx 2.33$

27. $\frac{11}{15}=0.7\bar{3}$

29. $3\sqrt{81}-8\sqrt{49}$

$=3(9)-8(7)$

$=27-56$

$=-29$

31. $(m^2-m-5)-(3m^2+2m-8)$

$=m^2-m-5-3m^2-2m+8$

$=-2m^2-3m+3$

33. $(2y-5)^2$
$=4y^2-10y-10y+25$
$=4y^2-20y+25$

35. $(h^5)^4=h^{20}$

37. $-7g^5(8g^4)=-56g^9$

39. $3x-3y=9$
$3y=3x-9$
$y=x-3$

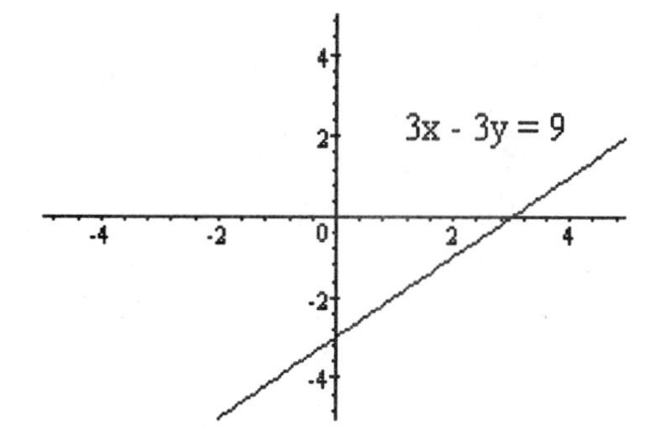

41.

Percent	*Decimal*	*Fraction*
29%	0.29	$\frac{29}{100}$
47.3%	0.473	$\frac{473}{1,000}$
87.5%	0.875	$\frac{875}{1,000}=\frac{7}{8}$

43. TIPPING
15% of \$75.18 is $0.15(75.18)=\$11.277$ so the total gratuity is \$11.28, and rounded up to the nearest dollar is \$12. Therefore the total to be paid is $\$75.18+\$12=\$87.18$.

45. SAVING ACCOUNT
$I=10,000(0.0725)(2)=\$1,450$

Section 8.1 Ratio

Vocabulary

1. A **ratio** is a quotient of two numbers or a quotient of two quantities with the same units.
3. When the price of candy is advertised at $1.75 per pound, we are told its unit **cost**.

Concepts

5. 15 and 24 have a common factor of 3.
7. To make a ratio of whole numbers multiply the numerator and denominator by 10.
9. Rewrite the denominator as 60 minutes.

Notation

11. $\frac{13}{9}$, 13 to 9, 13:9

Practice

13. 5 to 7 is $\frac{5}{7}$

15. 17 to 34 is $\frac{17}{34} = \frac{1}{2}$

17. $22:33 = \frac{22}{33} = \frac{2}{3}$

19. $1.5:2.4 = \frac{15}{24} = \frac{5}{8}$

21. $7 \text{ to } 24.5 = \frac{70}{245} = \frac{2}{7}$

23. $\frac{4 \text{ ounces}}{12 \text{ ounces}} = \frac{1}{3}$

25. $\frac{12 \text{ minutes}}{1 \text{ hour}} = \frac{12 \text{ minutes}}{60 \text{ minutes}} = \frac{1}{5}$

27. $\frac{3 \text{ days}}{1 \text{ week}} = \frac{3 \text{ days}}{7 \text{ days}} = \frac{3}{7}$

29. $\frac{18 \text{ months}}{2 \text{ years}} = \frac{18 \text{ months}}{24 \text{ months}} = \frac{3}{4}$

31. The total amount of this budget is $800+600+180+100+120=\$1,800$.

33. $\dfrac{\text{food}}{\text{total budget}}=\dfrac{600}{1800}=\dfrac{1}{3}$

35. The total amount of deductions is $875+1,250+1,750+4,375+500=\$8750$.

37. $\dfrac{\text{charitable contributions}}{\text{total deductions}}=\dfrac{1750}{8750}=\dfrac{1}{5}$

39. $\dfrac{64\text{ feet}}{6\text{ seconds}}=\dfrac{32\text{ feet}}{3\text{ seconds}}$

41. $\dfrac{84\text{ made}}{100\text{ attempts}}=\dfrac{21\text{ made}}{25\text{ attempts}}$

43. $\dfrac{3000\text{ students}}{16\text{ years}}=\dfrac{375\text{ students}}{2\text{ years}}$

45. $\dfrac{18\text{ beats}}{12\text{ measures}}=\dfrac{3\text{ beats}}{2\text{ measures}}$

47. 12 revolutions per minute

49. 1.5 errors per hour

51. 7 presents per child

53. 320 people per square mile

55. $\dfrac{\$3.50}{50\text{ ft}}=\$0.07$ per foot

57. $\dfrac{78\text{ cents}}{65\text{ ounces}}=1.2$ cents per ounce

59. $\dfrac{\$272}{4\text{ people}}=\$68$ per person

61. $\dfrac{\$4\text{ billion}}{5\text{ months}}=\$0.8$ billion per month

Applications

63. ART HISTORY
The ratio of the man's outstretched arms to his height is 1:1.

65. GEAR RATIO
The ratio of the number of teeth of the larger gear to the smaller gear is $\frac{18}{12}=\frac{3}{2}$.

67. COOKING
The ratio of sugar to milk is $\frac{\frac{2}{3}}{3\frac{1}{2}}$.

69. SOFTBALL
The rate of hits to at bats is $\frac{9 \text{ hits}}{33 \text{ at bats}}=\frac{3 \text{ hits}}{11 \text{ at-bats}}$.

71. CPR
The compressions to breaths rate was $\frac{125 \text{ compressions}}{50 \text{ breaths}}=\frac{5 \text{ compressions}}{2 \text{ breaths}}$.

73. AIRLINE COMPLAINTS
As a fraction of whole numbers $\frac{3.29 \text{ complaints}}{1,000 \text{ passengers}}=\frac{329 \text{ complaints}}{100,000 \text{ passengers}}$

75. FACULTY-STUDENT RATIO
$\frac{125 \text{ faculty}}{2,000 \text{ students}}=\frac{1 \text{ faculty}}{16 \text{ students}}$

77. UNIT COST
$\frac{\$32.13}{17 \text{ gallons}}=\$1.89/\text{gallon}$

79. UNIT COST
$\frac{84 \text{ cents}}{12 \text{ ounces}}=7 \text{ cents per ounce}$

81. COMPARISON SHOPPING

The six-ounce can unit price is $\dfrac{89 \text{ cents}}{6 \text{ ounce}} = 14.83 \text{ cents/ounce}$.

The eight-ounce can unit price is $\dfrac{119 \text{ cents}}{8 \text{ ounce}} = 14.875 \text{ cents/ounce}$.

The six-ounce can is a better buy.

83. COMPARISON SHOPPING

The 20-tablet product has a unit price of $\dfrac{\$4.29}{20 \text{ tablets}} = \$0.2145 \text{ per tablet}$.

The 50-tablet product has a unit price of $\dfrac{\$9.59}{50 \text{ tablets}} = \$0.1918 \text{ per tablet}$.

The 50-tablet product is a better buy.

85. COMPARING SPEEDS

The automobile rate of speed is $\dfrac{345 \text{ miles}}{6 \text{ hours}} = 57.5 \text{ mph}$.

The truck rate of speed is $\dfrac{376 \text{ miles}}{6.2 \text{ hours}} = 60.65 \text{ mph}$.

The truck is traveling at a faster rate.

87. EMPTYING A TANK

$\dfrac{11{,}880 \text{ gallons}}{27 \text{ minutes}} = 440 \text{ gallons/minute}$

89. AUTO TRAVEL

This car has traveled $35{,}071 - 34{,}746 = 325 \text{ miles}$.

The average rate of speed is $\dfrac{325 \text{ miles}}{5 \text{ hours}} = 65 \text{ mph}$.

91. GAS MILEAGE

Car one's gas mileage was $\dfrac{1{,}235 \text{ miles}}{51.3 \text{ gallons}} = 24.07 \text{ mpg}$.

Car two's gas mileage was $\dfrac{1{,}456 \text{ miles}}{55.78 \text{ gallons}} = 26.10 \text{ mpg}$.

The second automobile had better gas mileage.

Writing

93. Answers will vary.
95. Answers will vary.

Review

97. $3.05 + 17.17 + 25.317 = 45.537$

99. $13.2 + 25.07 \cdot 7.16 = 13.2 + 179.5012 = 192.7012$

101. $5 - 3\frac{1}{4} = \frac{20}{4} - \frac{13}{4} = \frac{7}{4} = 1\frac{3}{4}$

Section 8.2 Proportion

Vocabulary

1. A **proportion** is a statement that two ratios or rates are equal.
3. In $\frac{5}{6}=\frac{x}{18}$, the terms 6 and x are called the **means** of the proportion.

Concepts

5. The equation $\frac{a}{b}=\frac{c}{d}$ will be a proportion if the product *ad* is equal to the product *bc*.

7. a. $\frac{5}{8}=\frac{15}{24}$

   b. $\frac{\text{3 teacher's aides}}{\text{25 children}}=\frac{\text{12 teacher's aides}}{\text{100 children}}$

9. The following proportions could be used to solve this problem $\frac{15}{4}=\frac{300}{x}$ or $\frac{4}{15}=\frac{x}{300}$.

Notation

11.
$$12 \cdot 24 = 18x$$
$$288 = 18x$$
$$\frac{288}{18}=\frac{18x}{18}$$
$$16 = x$$

Practice

13. This is not a proportion. $\frac{9}{7}\neq\frac{81}{70}$
15. This is a proportion.
17. This is not a proportion. $\frac{9}{19}\neq\frac{38}{80}$
19. This is a proportion.
21. This is not a proportion. $\frac{\frac{2}{3}}{\frac{5}{8}}\neq\frac{\frac{4}{5}}{\frac{9}{16}}$
23. This is a proportion.

25. $\frac{2}{3}=\frac{x}{6}$

$12=3x$

$4=x$

27. $\frac{5}{10}=\frac{3}{c}$

$5c=30$

$c=6$

29. $\frac{6}{x}=\frac{8}{4}$

$24=8x$

$3=x$

31. $\frac{x}{3}=\frac{9}{3}$

$3x=27$

$x=9$

33. $\frac{x+1}{5}=\frac{3}{15}$

$15x+15=15$

$15x=0$

$x=0$

35. $\frac{x+3}{12}=\frac{-7}{6}$

$6x+18=-84$

$6x=-102$

$x=-17$

37. $\frac{4-x}{13}=\frac{11}{26}$

$104-26x=143$

$-26x=39$

$x=-\frac{39}{26}$

$x=-\frac{3}{2}$

39. $\dfrac{2x+1}{18}=\dfrac{14}{3}$

$$6x+3=252$$
$$6x=249$$
$$x=\frac{83}{2}$$

41. $\dfrac{4,000}{x}=\dfrac{3.2}{2.8}$

$$11,200=3.2x$$
$$3,500=x$$

43. $\dfrac{\frac{1}{2}}{\frac{1}{5}}=\dfrac{x}{2\frac{1}{4}}$

$$\frac{9}{8}=\frac{1}{5}x$$
$$\frac{45}{8}=x$$
$$5.625=x$$

Applications

45. SCHOOL LUNCHES

$$\frac{750}{x}=\frac{6}{1.75}$$
$$1,312.5=6x$$
$$\$218.75=x$$

47. GARDENING

$$\frac{3}{0.98}=\frac{36}{c}$$
$$3c=35.28$$
$$c=\$11.76$$

49. BUSINESS PERFORMANCE

The ratio of costs to revenues for 1999 is 2:1 and for 2000 is 2.5:1 so they are not the same.

51. MIXING PERFUME

$$\frac{3}{7}=\frac{d}{56}$$

$$168=7d$$

$24=d$ drops of pure essence

53. LAB WORK

$$\frac{195}{5}=\frac{x}{25}$$

$$4875=5x$$

$975=x$ red blood cells in the entire grid

55. MAKING COOKIES

$$\frac{1\frac{1}{4}}{3\frac{1}{2}}=\frac{f}{12}$$

$$\frac{7}{2}f=15$$

$$f=\frac{30}{7}$$

$f=4\frac{2}{7}$ cups of flour for 12 dozen cookies

57. COMPUTERS

$$\frac{15}{2.85}=\frac{100}{t}$$

$$15t=285$$

$t=19$ seconds for 100 calculations

59. FUEL CONSUMPTION

$$\frac{325}{25}=\frac{m}{17}$$

$$5{,}525=25m$$

$221=m$ miles on auxilary fuel cell

61. PAYCHECK

$$\frac{412}{40}=\frac{p}{30}$$

$$12{,}360=40p$$

$$\$309=p$$

63. BLUEPRINT

$$\frac{\frac{1}{4}}{1}=\frac{2\frac{1}{2}}{d}$$

$$\frac{1}{4}d=2\frac{1}{2}$$

$$\frac{1}{4}d=\frac{5}{2}$$

$d=10$ feet is the kitchen length

65. MODEL RAILROADING

$\frac{87}{1}=\frac{l}{\frac{3}{4}}$ (Note: convert 9 inches to feet, $\frac{9}{x}=\frac{12}{1}\rightarrow x=\frac{3}{4}$ ft)

$65.25=l$ is the length of the caboose measured in feet

67. MINIATURES

$\frac{13\frac{1}{3}}{1}=\frac{35}{w}$ (Note: convert 160 inches to feet, $\frac{160}{x}=\frac{12}{1}\rightarrow x=13\frac{1}{3}$ ft)

$$\frac{40}{3}w=35$$

$$w=\frac{21}{8}$$

$w=2\frac{5}{8}$ inches is the width of the model for a 35 foot wide carousel

Writing

69. Answers will vary.
71. Answers will vary.

Review

73. $\frac{9}{10}$ is 90%

75. $33\frac{1}{3}\%$ is $\frac{1}{3}$

77. $0.005(520) = 2.6$ (Note: $\frac{1}{2}\% = 0.50\% = 0.005$)

Section 8.3 American Units of Measurement

Vocabulary

1. Inches, feet and miles are examples of American units of **length**.
3. The value of any unit conversion factor is **1**.
5. Some examples of American units of **capacity** are cups, pints, quarts, and gallons.

Concepts

7. 12 in = 1 ft
9. 1 mi = 5,280 ft
11. 16 ounces = 1 pound
13. 1 cup = 8 fluid ounces
15. 2 pints = 1 quart
17. 1 day = 24 hours

19. From left to right the arrows point to $\frac{5}{8}$ inch, $1\frac{3}{4}$ inch and $2\frac{5}{16}$ inches.

21. a. $\dfrac{1 \text{ ton}}{2{,}000 \text{ lb}}$

    b. $\dfrac{2 \text{ pints}}{1 \text{ qt}}$

23. a. Length of the U.S. coastline matches with iv) 12,383 mi
    b. Height of a Barbie doll matches with i) 11 ½ inches
    c. Span of the Golden Gate Bridge matches with ii) 4,200 feet
    d. Width of a football field matches with iii) 53.5 yd

25. a. Amount of blood in an adult matches with iii) 5 qt
    b. Size of the Exxon Valdez oil spill in 1989 matches with iv) 10,080,000 gal
    c. Amount of nail polish in a bottle matches with i) ½ fluid oz
    d. Amount of flour to make 3-dozen cookies matches with ii) 2 cups

Notation

27.
$$\begin{aligned} 12 \text{ yd} &= 12 \text{ yd} \cdot \frac{36 \text{ in}}{1 \text{ yd}} \\ &= 12 \cdot 36 \text{ in} \\ &= 432 \text{ in} \end{aligned}$$

29. $12\text{ pt} = 12\text{ pt}\cdot\frac{1\text{ qt}}{2\text{ pt}}\cdot\frac{1\text{gallon}}{4\text{ qt}}$

$$= \frac{12\cdot 1\cdot 1}{2\cdot 4}\text{ gallons}$$

$$= 1.5\text{ gallons}$$

Practice

31. The width of dollar bill is approximately 2 5/8 inches.
33. The length of this page is approximately 10 ¾ inches.
35. 4 feet is 48 inches
37. $3\frac{1}{2}$ feet is 42 inches
39. 24 inches is 2 feet
41. 8 yards is 288 inches
43. 90 inches is $2\frac{1}{2}$ yards
45. 56 inches is $4\frac{2}{3}$ feet
47. 5 yards is 15 feet
49. 7 feet is $2\frac{1}{3}$ yards
51. 15,840 feet is 3 miles
53. ½ mile is 2,640 feet
55. 80 ounces is 5 pounds
57. 7,000 pounds is 3 ½ tons
59. 12.4 tons is 24,800 pounds
61. 3 quarts is 6 pints
63. 16 pints is 2 gallons
65. 32 fluid ounces is 2 pints
67. 240 minutes is 4 hours
69. 7,200 minutes is 5 days

Applications

71. THE GREAT PYRAMID

    450 feet is 150 yards

    $\frac{450}{3} = 150$

73. THE GREAT SPHINX

    240 feet is 2,880 inches

    $240(12) = 2,880$

75. THE SEARS TOWER

1,454 feet is approximately 0.28 miles

$$\frac{1,454}{5,280} \approx 0.27538$$

77. NFL RECORDS

35 miles is 61,600 yards

$$\frac{35(5,280)}{3} = 61,600$$

79. WEIGHT OF WATER

1 gallon is 16(8) = 128 ounces

81. HIPPOS

9,900 pounds is 4.95 tons

$$\frac{9,900}{2,000} = 4.95$$

83. BUYING PAINT

17 gallons is 68 quarts

$$17(4) = 68$$

85. SCHOOL LUNCHES

575 pints is $71\frac{7}{8}$ gallons

$$\frac{575}{8} = 71.875$$

87. CAMPING

2 ½ gallons is 320 ounces

$$\frac{5}{2}(128) = 320$$

89. SPACE TRAVEL

147 hours is $6\frac{1}{8}$ days

$$\frac{147}{24} = 6.125$$

Writing

91. Answers may vary

Review

93. 3,673.263 to the nearest hundred is 3,700
95. 3,673.263 to the nearest hundredth is 3,673.26
97. 0.100602 to the nearest thousandth is 0.101
99. 0.09999 to the nearest tenth is 0.1

Section 8.4 Metric Units of Measurement

Vocabulary

1. Deka means **tens.**
3. Kilo means **thousands.**
5. Centi means **hundredths.**
7. Meters, grams, and liters are units of measurement in the **metric** system.

Concepts

9. From left to right the measurement arrows point to 1 cm, 3 cm, and 6 cm.

11. a. $\frac{1 \text{ km}}{1{,}000 \text{ m}}$

    b. $\frac{100 \text{cg}}{1 \text{ g}}$

    c. $\frac{1{,}000 \text{ ml}}{1 \text{ liter}}$

13. a. Thickness of a phone book matches with iii) 6 cm
    b. Length of the Amazon River matches with i) 6,275 km
    c. Height of a soccer goal matches with ii) 2 m

15. a. Amount of blood in an adult matches with ii) 6 L
    b. Cola in an aluminum can matches with iii) 355 mL
    c. Kuwait's daily production of crude oil matches with i) 290,000 kL

17. 1 dekameter = 10 meters

19. 1 centimeter = $\frac{1}{100}$ meter

21. 1 millimeter = $\frac{1}{1{,}000}$ meters

23. 1 gram = 1,000 milligrams
25. 1 kilogram = 1,000 grams
27. 1 liter = 1,000 cubic centimeters

29. 1 centiliter = $\frac{1}{100}$ liter

31. 100 liters = 1 hectoliter

Notation

33. $20 \text{ cm} = 20 \text{ cm} \cdot \frac{1m}{100 \text{ cm}}$

$= \frac{20}{100} \text{ m}$

$= 0.2 \text{ m}$

35. $2 \text{ km} = 2 \text{ km} \cdot \frac{1{,}000 \text{ m}}{1 \text{ km}} \cdot \frac{10 \text{ dm}}{1 \text{ m}}$

$= 2 \cdot 1{,}000 \cdot 10 \text{ dm}$

$= 20{,}000 \text{ dm}$

Practice

37. The length of a dollar bill is approximately 156 mm.
39. The length of this page is approximately 28 cm.
41. 3 m = 300 cm
43. 5.7 m = 570 cm
45. 0.31 dm = 3.1 cm
47. 76.8 hm = 7,680,000 mm
49. 4.72 cm = 0.472 dm
51. 453.2 cm = 4.532 m
53. 0.325 dm = 0.0325 m
55. 3.75 cm = 37.5 mm
57. 0.125 m = 125 mm
59. 675 dam = 675,000 cm
61. 638.3 m = 6.383 hm
63. 6.3 mm = 0.63 cm
65. 695 dm = 69.5 m
67. 5,689 m = 5.689 km
69. 576.2 mm = 5.762 dm
71. 6.45 dm = 0.000645 km
73. 658.23 m = 0.65823 km
75. 3 g = 3,000 mg
77. 2 kg = 2,000 g
79. 1,000 kg = 1,000,000 g
81. 500 mg = 0.5 g
83. 3 kL = 3,000 L
85. 500 cL = 5,000 mL
87. 10 mL = 10 cc

Applications

89. SPEED SKATING
    500 m = 0.5 km
    1,000 m = 1 km
    1,500 m = 1.5 km
    5,000 m = 5 km
    10,000 m = 10 km

91. HEALTH CARE
    120 mm = 12 cm
    80 mm = 8 cm

93. WEIGHT OF A BABY
    4 kg = 400,000 cg

95. CONTAINERS
    $2(2\text{L}) = 4\text{L} = 40$ dL

97. BUYING OLIVES
    1,000 g = 1 kg so 4 bottles of olives is 1,136 grams or 1.136 kg.

99. MEDICINE
    In the bottle there are $60(50) = 3{,}000$ mg $= 3$g of active ingredient.

Writing

101. Answers may vary
103. Answers may vary

Review

105. $0.07(342.72) = 23.9904 = \$23.99$

107. $\$32.16 = 0.08x$
$\$402 = x$

109.

$$3\frac{1}{7}+2\frac{1}{2}\bullet 3\frac{1}{3}=\frac{22}{7}+\frac{5}{2}\bullet\frac{10}{3}$$
$$=\frac{22}{7}+\frac{25}{3}$$
$$=\frac{66}{21}+\frac{175}{21}$$
$$=\frac{241}{21}$$
$$=11\frac{10}{21}$$

Section 8.5 Converting between American and Metric Units

Vocabulary

1. In the American system, temperatures are measured in degrees **Fahrenheit**.

Concepts

3. a. A meter is longer than a yard.
   b. A meter is longer than a foot.
   c. An inch is longer than a centimeter.
   d. A mile is longer than a kilometer.

5. a. A liter is greater than a pint.
   b. A liter is greater than a quart.
   c. A gallon is greater than a liter.

Notation

7. $4{,}500 \text{ ft} = 4{,}500 \text{ ft} \, (0.3048 \text{m/ft})$
   $= 1371.6 \text{ m}$
   $= 1.3716 \text{ km}$

9. $8 \text{ L} = 8 \text{ L} \, (0.264 \text{gal/L})$
   $= 2.112 \text{ gal}$

Practice

11. 3 ft = 91.4 cm
13. 3.75 m = 147.6 in
15. 12 km = 39,372 ft
17. 5,000 in = 127 m
19. 37 oz = 1 kg
21. 25 lb = 11,350 g
23. 0.5 kg = 17.6 oz
25. 17 g = 0.6 oz
27. 3 fl oz = 0.1 L
29. 7.2 L = 243.4 fl oz
31. 0.75 qt = 710 mL
33. 500 mL = 0.5 qt
35. 50^0 F = $10°$ C
37. 50^0 C = $122°$ F
39. -10^0 C = $14°$ F
41. -5^0 F = $-20.6°$ C

Applications

43. THE MIDDLE EAST
To the nearest mile, 8 km is 8(0.6214) = 5 miles.

45. CHEETAH
112 km/hr = 112(0.6214) = 170 mph

47. MOUNT WASHINGTON
6,288 ft = 6,288(0.3048/1000) = 1.9 km

49. HAIR GROWTH
¾ inch per month is approximately ¾(2.54) = 1.9 cm per month

51. WEIGHTLIFTING
Ms. Steenrod's weight class was 82.5/0.454 = 181.7 pounds and her bench press was 132.5/0.454 = 291.9 pounds. Mr. Magruder's weight class was 110/0.454 = 242.3 pounds and his bench press was 270/0.454 = 594.7 pounds.

53. OUNCES AND FLUID OUNCES
a. 8 oz = 8(28.35) = 226.8 g
b. 8 fluid oz = 8/33.8 = 0.24 L

55. POSTAL REGULATIONS
32 kg = 32(2.2) = 70.4 pounds therefore it cannot be sent by priority mail

57. COMPARISON SHOPPING

Using 1 qt = 0.946 L, then 3 qt = 2.838 L, the price per liter is $\dfrac{\$\ 4.50}{2.838 \text{ L}} = \$\ 1.59$ per liter.

The price per liter for the two liter root beer is $\dfrac{\$\ 3.60}{2 \text{ L}} = \$1.80$ per liter.

The 3 quarts is a better buy.

59. HOT SPRINGS

$$C = \frac{5}{9}(F - 32)$$

$$C = \frac{5}{9}(143 - 32)$$

$$C = 61.7^0$$

61. TAKING A SHOWER

The temperature to choose is 28^0 C.

$$F=\frac{9}{5}C+32 \qquad F=\frac{9}{5}C+32 \qquad F=\frac{9}{5}C+32$$

$$F=\frac{9}{5}(28)+32 \qquad F=\frac{9}{5}(15)+32 \qquad F=\frac{9}{5}(50)+32$$

$$F=82.4^0 \qquad F=59^0 \qquad F=122^0$$

63. SNOWY WEATHER

It might snow at -5^0 C and 0^0 C.

$$F=\frac{9}{5}C+32 \qquad F=\frac{9}{5}C+32 \qquad F=\frac{9}{5}C+32$$

$$F=\frac{9}{5}(0)+32 \qquad F=\frac{9}{5}(-5)+32 \qquad F=\frac{9}{5}(10)+32$$

$$F=32^0 \qquad F=23^0 \qquad F=50^0$$

Writing

65. Answers may vary
67. Answers may vary

Review

69. $6y+7-y-3=5y+4$
71. $-3(x-4)-2(2x+6)=-3x+12-4x-12=-7x$
73. $x \bullet x \bullet x = x^3$
75. $3b(5b)=15b^2$

Chapter 8 Key Concepts

1. TEACHER'S AIDES

   Step 1:

   Let x = the number of teacher's aides needed to supervise 75 children.

   15 children are to 2 aides as 75 children are to x aides

   $$\frac{15}{2}=\frac{75}{x}$$

   Step 2:

   $$\frac{15}{2}=\frac{75}{x}$$

   $$15 \cdot x = 2 \cdot 75$$

   $$15x = 150$$

   $$\frac{15x}{15}=\frac{150}{15}$$

   $$x = 10$$

   Step 3:

   $$15 \cdot 10 = 150$$

   $$2 \cdot 75 = 150$$

3. MOTION PICTURES

   Convert 120 minutes to seconds, $120 \cdot 60 = 7,200$.

   $$\frac{2}{3}=\frac{7,200}{x}$$

   $$2x = 21,600$$

   $$x = 10,800$$

   There are 10,800 feet of film.

Chapter 8 Review

Section 8.1 Ratio

1. a. $\frac{4}{12}=\frac{1}{3}$

   b. Convert 2 pounds to ounces, 2(16) = 32 ounces, then $\frac{8}{32}=\frac{1}{4}$.

   c. 21:14 is $\frac{3}{2}$

   d. 24 to 36 is $\frac{2}{3}$

3. PAY SCALE

   $\frac{\$\ 333.25}{43}=\$\ 7.75$ per hour

Section 8.2 Proportion

5. a. 75 is the fourth term

   b. 15 is the second term

7. a. These are proportional since $5(36)=9(20)$

   b. These are not proportional since $7(54)\neq 13(29)$

9. PICKUP TRUCK

   $\frac{2}{35}=\frac{11}{m}$

   $2m=385$

   $m=192\frac{1}{2}$ miles on 11 gallons

11. SCALE DRAWING

   $\frac{\frac{1}{8}}{1}=\frac{1\frac{1}{2}}{x}$

   $\frac{1}{8}x=\frac{3}{2}$

   $x=12$ ft

   The kitchen is 12 feet long.

Section 8.3 American Units of Measure

13. $\dfrac{5{,}280\text{ ft}}{1\text{ mi}};\dfrac{1\text{ mi}}{5{,}280\text{ ft}}$

15. a. 32 oz = 32/16 = 2 lbs
    b. 17.2 lb = 17.2(16) = 275.2 oz
    c. 3 tons = 3(2000)(16) = 96,000 oz
    d. 4,500 lb = 4,500/2000 = 2.25 tons

17. a. 20 min = 20(60) = 1,200 sec
    b. 900 sec = 900/60 = 15 min
    c. 200 hr = 200/24 = 8 1/3 days
    d. 6 hr = 6(60) = 360 min
    e. 4.5 days = 4.5(24) = 108 hr
    f. 1 day = 1(24)(60)(60) = 86,400 sec

19. BOTTLING
    50 gallons of wine is 50(4) = 200 quarts. If each magnum holds 2 quarts, then 200/2 = 100 magnums are needed to hold 50 gallons of wine.

Section 8.4 Metric Units of Measure

21. $\dfrac{1{,}000\text{ m}}{1\text{ km}};\dfrac{1\text{ km}}{1{,}000\text{ m}}$

23. a. 7 cg = (7/100)(1,000) = 70 mg
    b. 800 cg = 800/100 = 8 g
    c. 5,425 g = 5,425/1,000 = 5.425 kg
    d. 5,425 g = 5,425(1,000) = 5,425,000 mg
    e. 7,500 mg = 7,500/1,000 = 7.5 g
    f. 5,000 cg = $5{,}000 \div 100 \div 1{,}000 = 0.05$ kg

25. a. 150 cL = 150/100 = 1.5 L
    b. 3,250 L = 3,250/1,000 = 3.25 kL
    c. 1 hL = 1(100)(10) = 1,000 dL
    d. 400 mL = (400/1,000)(100) = 40 cL
    e. 2 kL = 2(1,000/100) = 20 hL
    f. 4 dL = (4/10)(1,000) = 400 mL

Section 8.5 Converting Between American and Metric Units

27. SWIMMING

 50 m = 50(3.2808) = 164.04 ft

29. WESTERN SETTLERS

 To the nearest kilometer, 1,930 miles is 1,930/0.6214 = 3,106 km.

31. a. 30 oz = 30(28.35) = 850.5 g
 b. 15 kg = 15(2.2) = 33 lb
 c. 25 lb = 25(0.454)(1,000) = 11,000 g
 d. 2,000 lb = 2,000/2.2 = 910 kg

33. BOTTLED WATER

 0.5 L = 0.5(33.8) = 16.9 oz, so the LaCroix bottle contains more water.

35. $C = \frac{5(77) - 160}{9}$

 $C = 25^0$

Chapter 8 Test

1. The ratio of 6 feet to 8 feet is $\frac{6}{8}=\frac{3}{4}$.

3. COMPARISON SHOPPING

   The 2 lb can costs $\frac{3.38}{2}=\$\,1.69/\text{pound}$ and the 5 lb can costs $\frac{8.50}{5}=\$1.70/\text{pound}$.

   The 2 pound can is a better buy.

5. CHECKERS

   $\frac{1}{1}$; 1:1 and 1 to 1

7. This is a proportion since $2.2(2.8)=3.5(1.76)$.

9. $$\frac{x}{3}=\frac{35}{7}$$
   $$7x=105$$
   $$x=15$$

11. $$\frac{2x+3}{5}=\frac{5}{1}$$
    $$2x+3=25$$
    $$2x=22$$
    $$x=11$$

13. SHOPPING

    $$\frac{2.79}{13}=\frac{c}{16}$$
    $$44.64=13c$$
    $$\$3.43=c$$

    16 oz should cost \$3.43

15. 180 in = 180/12 = 15 ft

17. 10 lb = 10(16) = 160 oz

19. 1 gal = 1(4)(2)(2)(8) = 128 fl oz

21. The carton on the left is the 1-liter carton since 1L is greater than 1 quart.

23. The gram weight is on the right side since 1 gm weighs less than 1 ounce.

25. There are $5(100) = 500$ centimeters in 5 meters.

27. $70 \text{ L} = 70(1{,}000) = 70{,}000 \text{ mL}$

29. The 100-yard race is longer than an 80-meter race since $100 \text{ yd} = 100(0.9144) = 91.44 \text{ m}$.

31. COMPARISON SHOPPING
    The two quart bottle is equivalent to $2(0.946) = 1.892$ liters, so the per liter cost is $(1.73/1.892) = \$0.91$, thus the one liter bottle is a better buy.

33. Answers may vary

Chapters 1-8 Cumulative Review Exercises

1. 64,502 is six ten-thousands + four thousands + five hundreds + zero tens + 2 ones

3. ENLISTMENT

Air Force	$-1732 = 32,068 - 33,800$
Army	$-6,290 = 68,210 - 74,500$
Marines	$35 = 39,521 - 39,486$
Navy	$71 = 52,595 - 52,524$

5. PROFESSIONAL GOLF
   Wood's margin of victory was 15 shots, $3-(-12)=15$

7. $2+3[5(-6)-1(1-10)]$
   $=2+3[-30-1(-9)]$
   $=2+3[-21]$
   $=2-63$
   $=-61$

9. There are $60h$ minutes in h hours.

11. $7+2x=2-(4x+7)$
    $7+2x=2-4x-7$
    $7+2x=-4x-5$
    $6x=-12$
    $x=-\dfrac{12}{6}$
    $x=-2$

13. PHONE BOOKS
    Let b = the number of office buildings
    $500-5b=105$
    $500-105=5b$
    $395=5b$
    $79=b$
    He delivered to 79 office buildings.

15. $\dfrac{16}{20}=\dfrac{4}{5}$

17. $-\dfrac{7}{8h}\div\dfrac{7}{8}=-\dfrac{7}{8h}\bullet\dfrac{8}{7}=-\dfrac{1}{h}$

19. MOTORS

$$1\frac{1}{2}-\frac{3}{4}=\frac{6}{4}-\frac{3}{4}=\frac{3}{4}\text{ hp}$$

21. $\frac{2}{5}y+1=\frac{1}{3}+y$

$$1-\frac{1}{3}=y-\frac{2}{5}y$$

$$\frac{2}{3}=\frac{3}{5}y$$

$$\frac{10}{9}=y$$

23. Let $x=-0.3$

$$=\frac{6.7-(-0.3)^2+1.6}{-(-0.3)^3}$$

$$=\frac{8.21}{0.027}$$

$$=304.07$$

25. $6(y-1.1)+3.2=-1+3y$

$$6y-6.6+3.2=-1+3y$$

$$6y-3.4=-1+3y$$

$$3y=2.4$$

$$y=0.8$$

27. $A(-4,-3), B(1.5,1.5), C(-3,0), D(0,3\frac{1}{2})$

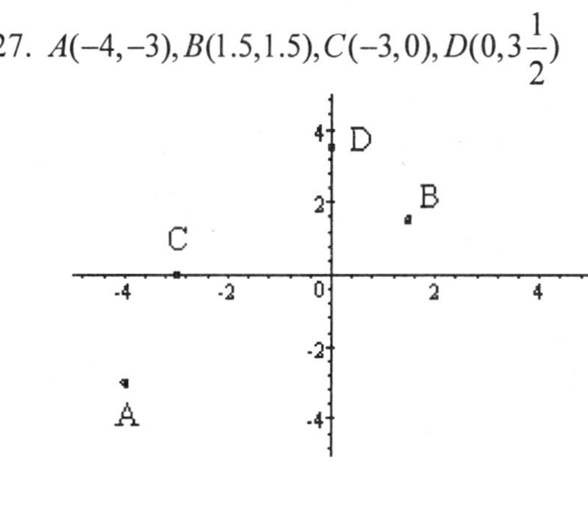

29. $(5x^2 - 8x + 1) - (3x^2 - 2x + 3)$

$= 5x^2 - 8x + 1 - 3x^2 + 2x - 3$

$= 2x^2 - 6x - 2$

31. a. $s^6 \bullet s^7 = s^{13}$

b. $(s^6)^7 = s^{42}$

c. $(3a^2b^4)^3 = 27a^6b^{12}$

d. $-w^5(8w^3) = -8w^8$

33. The simple interest formula is I = Prt.

35. GUITAR SALES

The regular price is \$299.99 + \$128 = \$427.99

The percent savings is $1 - \dfrac{299.99}{427.99} \approx 0.30$ or approximately 30%.

37. SURVIVAL GUIDE

a. 40 days = 40(24) = 960 hr

b. 3 days = 3(24)(60) = 4,320 min

c. 8 min = 8(60) = 480 sec

39. 2.4 m = 2.4(1,000) = 2,400 mm

41. a. A one-gallon bottle holds more than a two-liter bottle since 1 gal = 3.785L.

b. A meter stick is longer than a yardstick since 1 m = 1.0936 yd.

Section 9.1 Some Basic Definitions

Vocabulary

1. A line **segment** has two endpoints.
3. A **midpoint** divides a line segment into two parts of equal length.
5. A **protractor** is used to measure angles.
7. A **right** angle measures 90^0.
9. The measure of a straight angle is **180^0**.
11. The sum of two **supplementary** angles is 180^0.

Concepts

13. $\overrightarrow{GF}$ has point G as its endpoint is a **true** statement.
15. Line CD has three endpoints is a **false** statement. It has 2.
17. $m(\angle AGC) = m(\angle BGD)$ is a **true** statement.
19. $\angle FGB \cong \angle EGA$ is a **true** statement.
21. $\angle AGC$ is an **acute** angle
23. $\angle FGD$ is an **obtuse** angle.
25. $\angle BGE$ is a **right** angle.
27. $\angle DGC$ is a **straight** angle.
29. $\angle AGF$ and $\angle DGC$ are vertical angles is a **true** statement.
31. $m(\angle AGB) = m(\angle BGC)$ is a **false** statement.
33. $\angle 1$ and $\angle 2$ **are** congruent angles.
35. $\angle AGB$ and $\angle DGE$ **are** congruent angles.
37. $\angle AGF$ and $\angle FGE$ are **not** congruent angles.
39. $\angle 1$ and $\angle CGD$ are adjacent angles is a **true** statement.
41. $\angle FGA$ and $\angle AGC$ are supplementary angles is a **true** statement.
43. $\angle AGF$ and $\angle 2$ are complementary angles is a **true** statement.
45. $\angle EGD$ and $\angle DGB$ are supplementary angles is a **true** statement.

Notation

47. The symbol $\angle$ means **angle.**
49. The symbol $\overrightarrow{AB}$ is read as "**ray AB.**"

Practice

51. $\overline{AC}$ has length **3**.
53. $\overline{CE}$ has length **3**.
55. $\overline{CD}$ has length **1**.
57. The midpoint of $\overline{AD}$ is B.
59. The measure of this angle is 40^0.

61. The measure of this angle is 135^0.
63. The measure of angle x is $55^0 - 45^0 = 10^0$.
65. The measure of angle x is $50^0 - 22.5^0 = 27.5^0$.
67. Since these are vertical angles, $2x = x + 30 \to x = 30^0$.
69. Since these are vertical angles, $4x + 15 = 7x - 60$.

$$75 = 3x$$

$$25^0 = x$$

71. The complement of a 30^0 angle is 60^0 since $30 + x = 90$.
73. The supplement of a 105^0 angle is 75^0 since $105 + x = 180$.
75. $\angle 4$ is the supplement of $\angle 1$, therefore $\angle 4 = 180 - 50 = 130^0$.
77. $m(\angle 1) + m(\angle 2) + m(\angle 3) = 50 + 130 + 50 = 230^0$, since $\angle 2 = \angle 4$ by vertical angles.
79. $\angle 1 = 100^0$ by vertical angles.
81. Since we are told that $\angle 3 \cong \angle 4$, $\angle 3 + \angle 4 = 80^0$ since $\angle 1 = 100^0$, therefore $\angle 3 = 40^0$.

Applications

83. BASEBALL

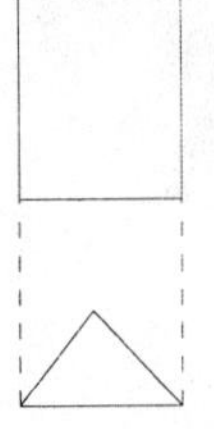

85. SYNTHESIZER
x is the supplement of 115^0, so $x = 180 - 115 = 65^0$
y is a vertical angle to 115^0, so $y = 115^0$.

87. GARDENING
The angle of the handle to the ground is the supplement of 150^0, or $180 - 150 = 30^0$.

Writing

89. Answers will vary.
91. Answers will vary.
93. Answers will vary.

Review

95. $2^4 = 2 \bullet 2 \bullet 2 \bullet 2 = 16$

97. $\frac{3}{4} - \frac{1}{8} - \frac{1}{3} = \frac{18}{24} - \frac{3}{24} - \frac{8}{24} = \frac{7}{24}$

99.

x	$y = 2x - 5$
-1	-8
0	-5
1	-3

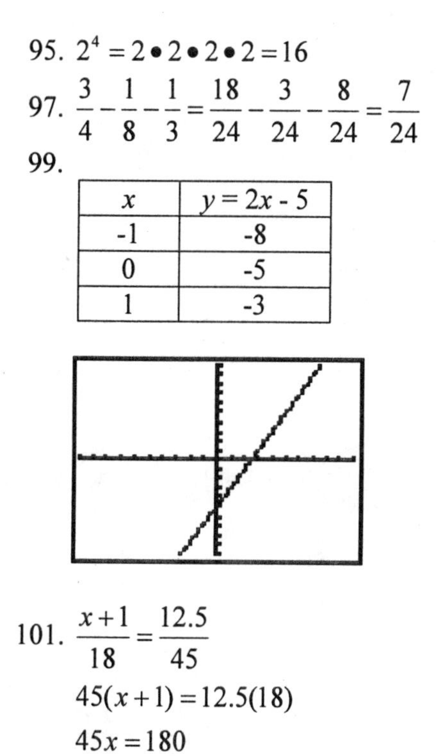

101. $\frac{x+1}{18} = \frac{12.5}{45}$

$45(x+1) = 12.5(18)$

$45x = 180$

$x = 4$

Section 9.2 Parallel and Perpendicular Lines

Vocabulary

1. Two lines in the same plane are **coplanar**.
3. If two lines intersect and form right angles, they are **perpendicular**.
5. In illustration 1, $\angle 4$ and $\angle 6$ are **alternate** interior angles.

Concepts

7. $\angle 4$ and $\angle 6$, $\angle 3$ and $\angle 5$ are pairs of alternating interior angles.
9. $\angle 3, \angle 4, \angle 5, \angle 6$ are each interior angles.
11. l_1 and l_3 are parallel lines.

Notation

13. The symbol $\urcorner$ represents **a right angle**.

15. The symbol $\perp$ is read as "**is perpendicular to**."

Practice

17. $m(\angle 1) = 130^0; m(\angle 2) = 180 - 30 = 50^0; m(\angle 3) = 50^0; m(\angle 5) = 130^0; m(\angle 6) = 50^0;$
$m(\angle 7) = 50^0; m(\angle 8) = 130^0$

19. $m(\angle A) = 50^0; m(\angle 1) = 85^0; m(\angle 2) = 45^0; m(\angle 3) = 180 - 45 = 135^0$

21. $6x - 10 = 5x$
$x = 10$

23. $2x + 10 + 4x - 10 = 180$
$6x = 180$
$x = 30^0$

25. $x + 3x + 20 = 180$
$4x = 160$
$x = 40^0$

27. $9x - 38 = 6x - 2$
$3x = 36$
$x = 12^0$

Applications

29. CONSTRUCTING PYRAMIDS
    If the stones are level the plum bob string should pass through the midpoint of the crossbar of the A frame.

31. LOGO
    The perpendicular lines form a cross-hair in the center of the logo.

33. HANGING WALLPAPER
    Answers will vary.

Writing

35. Answers will vary.
37. Answers will vary.
39. Answers will vary.

Review

41. 60% of 120 is $(0.60)(120) = 72$

43. $\frac{225}{500} = 0.45$, 225 is 45% of 500.

45. Every whole number is an integer.

47. $\frac{4}{12} = \frac{1}{3}$

Section 9.3 Polygons

Vocabulary

1. A **regular** polygon has sides that are all the same length and angles that all have the same measure.
3. A **hexagon** is a polygon with six sides.
5. An eight-sided polygon is an **octagon**.
7. A triangle with three sides of equal length is called an **equilateral** triangle.
9. The longest side of a right triangle is the **hypotenuse**.
11. A **parallelogram** with a right angle is a rectangle.
13. A **rhombus** is a parallelogram with four sides of equal length.
15. The legs of an **isosceles** trapezoid have the same length.

Concepts

17. 4, quadrilateral, 4
19. 3, triangle, 3
21. 5, pentagon, 5
23. 6, hexagon, 6
25. This is a scalene triangle.
27. This is a right triangle.
29. This is an equilateral triangle.
31. This is an isosceles triangle.
33. This is a square.
35. This is a rhombus.
37. This is a rectangle.
39. This is a trapezoid.

Notation

41. The symbol $\triangle$ means **triangle**.

Practice

43. $m(\angle C) = 180 - 60 - 30 = 90^0$
45. $m(\angle C) = 180 - 35 - 100 = 45^0$
47. $m(\angle C) = 180 - 25.5 - 63.8 = 90.7^0$
49. $m(\angle 1) = 90 - 60 = 30^0$
51. $m(\angle 2) = 90 - 30 = 60^0$
53. The sum of the angle measures of a hexagon is $180(6-2) = 720^0$.
55. The sum of the angle measures of a decagon is $180(10-2) = 1440^0$.

57. $180(n-2)=900^0$

$$n-2=5$$
$$n=7$$

59. $180(n-2)=2160$

$$n-2=12$$
$$n=14$$

Applications

61. Answers will vary.
63. Answers will vary.

65. POLYGONS IN NATURE
The lemon appears as an octagon.
The chili pepper appears as a triangle.
The apple appears as a pentagon.

67. CHEMISTRY
Pentagons and hexagons are used to represent methylprednisolone.

69. EASEL
The two outside legs form the beginning of an isosceles triangle.

Writing

71. Answers will vary.

Review

73. 20% of 110 is $(0.20)(110)=22$

75. $\dfrac{80}{200}=0.40$, 80 is 40% of 200.

77. $0.85\div 2(0.25)=0.425(0.25)=0.10625$

Section 9.4 Properties of Triangles

Vocabulary

1. **Congruent** triangles are the same size and the same shape.
3. If two triangles are **similar** they have the same shape.

Concepts

5. This is a true statement.
7. This is a false statement.
9. This is a true statement.
11. These are congruent triangles.
13. These are similar triangles.
15. For a right triangle, a and b represent the legs of the triangle and c represents the hypotenuse.
17. To find c, we must find a number that, when squared is **25**. Since c represents a positive number, we need only find the positive **square root** of 25 to get c.

$$c^2 = 25$$

$$c = \sqrt{25} = 5$$

Notation

19. The symbol $\cong$ is read as "**is congruent to.**"

Practice

21. $\overline{AC}$ corresponds to $\overline{DF}$

$\overline{DE}$ corresponds to $\overline{AB}$

$\overline{BC}$ corresponds to $\overline{EF}$

$\angle A$ corresponds to $\angle D$

$\angle E$ corresponds to $\angle B$

$\angle F$ corresponds to $\angle C$

23. These are congruent triangles by SSS.
25. These are not necessarily congruent.
27. These are congruent triangles by SSS.
29. These are congruent triangles by SAS.
31. $x = 6$ cm
33. $x = 50^0$
35. These are similar triangles.
37. Using $c^2 = a^2 + b^2$, $c = \sqrt{3^2 + 4^2} : c = 5$

39. Using $c^2 = a^2 + b^2$, $b = \sqrt{c^2 - a^2} : b = \sqrt{17^2 - 15^2} : b = 8$

41. Using $c^2 = a^2 + b^2$, $b = \sqrt{c^2 - a^2} : b = \sqrt{9^2 - 5^2} : a = \sqrt{56}$

43. This is a right triangle since $17^2 = 15^2 + 8^2$

45. This is not a right triangle since $26^2 \neq 7^2 + 24^2$.

Applications

47. HEIGHT OF A TREE

$\frac{h}{6} = \frac{24}{4} : h = 36$

The height of the tree is 36 feet.

49. WIDTH OF A RIVER

$\frac{w}{20} = \frac{74}{25} : w \approx 59.2$

The river is approximately 59 feet wide.

51. FLIGHT PATH

Let d represent the amount of descent:

$\frac{d}{1200} = \frac{5}{1.5} : d = 4000$

In 5 miles the planes losses 4000 feet of altitude.

53. ADJUSTING A LADDER

Use Pythagorean's theorem with the ladder representing the hypotenuse:

$20^2 = 16^2 + b^2 : b = \sqrt{20^2 - 16^2} : b = 12$

The ladder is 12 feet from the wall.

55. PICTURE FRAME

$c^2 = 15^2 + 20^2 : w = \sqrt{625} : c = 25$

If the sides of the frame form a right angle then the frame maker should read 25 inches on the yardstick.

57. BASEBALL

The distance from home plate to second base is the hypotenuse:

$d^2 = 90^2 + 90^2 : d = \sqrt{16200} : d \approx 127.3$

The distance is approximately 127.3 feet.

Writing

59. Answers will vary.

Review

61. Estimate would be $\frac{1(4)}{3} = 1\frac{1}{3}$; Actual answer is $\frac{0.95(3.89)}{2.997} \approx 1.233$

63. Estimate would be 0.3(60) = 18; Actual answer is $0.32(60) = 19.2$

65. Estimate would be 0.5(18) = 9; Actual answer is $0.495(18.1) = 8.9595$

Section 9.5 Perimeters and Areas of Polygons

Vocabulary

1. The distance around a polygon is called the **perimeter**.
3. The measure of the surface enclosed by a polygon is called its **area**.
5. The area of a polygon is measured in **square** units.

Concepts

7. Two possible rectangles: length of 15 inches and width of 5 inches or length of 16 inches and width of 4 inches.
9. A square with area of 25 m^2 has side length 5 m.
11. A parallelogram with an area of 15 yd^2 may have a base of 5 yd and a height of 3 yd.
13. One of the rectangles of this figure may have a length of 5 ft and a width of 4 ft and the other rectangle may have a length of 20 ft and a width of 3 ft.

Notation

15. The formula for the perimeter of a square is $P = 4s$.
17. The symbol 1 in^2 means one square inch.
19. The formula for the area of a square is $A = s^2$.
21. The formula $A = \frac{1}{2}bh$ gives the area of a triangle.

Practice

23. $P = 4(8) = 32$ inches

25. $P = 6 + 4 + 2 + 2 + 2 + 4 + 6 + 10 = 36$ inches

27. $P = 7 + 6 + 6 + 8 + 10 = 37$ cm

29. $P = 21 + 32 + 32 = 85$ cm

31. $\frac{85}{3} = 28\frac{1}{3}$

33. $A = 4^2 = 16 \text{ cm}^2$

35. $A = 4(15) = 60 \text{ cm}^2$

37. $A = \frac{1}{2}(5)(10) = 25 \text{ in}^2$

39. $A = \frac{1}{2}(13)(9 + 17) = 169 \text{ mm}^2$

41. $A=\frac{1}{2}(4)(8)+8^2=80\ \text{m}^2$

43. $A=10^2-\frac{1}{2}(5)(10)=75\ \text{yd}^2$

45. $A=6(14)-3^2=75\ \text{m}^2$

47. There are $12^2=144$ square inches in 1 square foot.

Applications

49. FENCING A YARD
The perimeter of the yard is $P=2(110)+2(85)=390$ ft.
The cost of the fence is $C=390(12.50)=\$4,875$.

51. PLANTING A SCREEN
The perimeter of her yard to be planted is $P=2(70)+100=240$ feet.
Planting a tree every 3 feet results in $\frac{240}{3}=80$ trees plus the first tree planted for a total of 81 trees.

53. BUYING A FLOOR
Since there are 9 square feet in a square yard the cost of the ceramic tile is $C=3.75(9)=\$33.75$ per square yard, therefore the linoleum is more expensive.

55. CARPETING A ROOM
The area of this room is $A=24(15)=360$ square feet or $\frac{360}{9}=40$ square yards. The cost to carpet this room is $C=40(30)=\$1,200$.

57. TILING A FLOOR
The area of this room is $A=14(20)=280$ square feet. At a cost of \$1.29 per square foot, it will cost $C=1.29(280)=\$361.20$ to tile this floor.

59. MAKING A SAIL
Since the nylon is priced by the square yard, convert the dimensions of the sail to yards. The dimensions of the sail are 4 yards by 8 yards. The area of this triangular sail is $A=\frac{1}{2}(4)(8)=16$ square yards. The cost to make this sail is $C=16(12)=\$192$.

61. GEOGRAPHY

The approximate area of Nevada is $A = \frac{1}{2}(315)(505+205) = 111,825$ square miles.

63. CARPENTRY

Assuming the drywall is hung horizontally, each long side will need $\frac{(48)(12)}{(4)(8)} = 18$ sheets, for a total of 36 sheets. The two ends of the barn will have the drywall hung vertically and each end will need $\frac{20(12)}{(4)(8)} = 7.5$ sheets, for a total of 15 sheets of drywall. The total is 51 sheets.

65. DRIVING SAFETY

Spot one has length of 20 feet and width of 10 feet, for total area of 200 square feet. Spot two has base dimensions of 20 feet and 16 feet with height of 10 feet, thus the area is 180 square feet. Spot three is triangular with a height of 28 feet and a base of 28 feet for an area of 392 square feet.

Writing

67. Answers will vary

Review

69. $\frac{3}{4}+\frac{2}{3}=\frac{9}{12}+\frac{8}{12}=1\frac{5}{12}$

71. $3\frac{3}{4}+2\frac{1}{3}=3\frac{9}{12}+2\frac{4}{12}=5+1\frac{1}{12}=6\frac{1}{12}$

73. $7\frac{1}{2}\div 5\frac{2}{5}=\frac{15}{2}\left(\frac{5}{27}\right)=\frac{25}{18}=1\frac{7}{18}$

Section 9.6 Circles

Vocabulary

1. A segment drawn from the center of a circle to a point on the circle is called a **radius.**
3. A **diameter** is a chord that passes through the center of a circle.
5. An arc that is shorter than a semicircle is called a **minor** arc.
7. The distance around a circle is called its **circumference**.

Concepts

9. Radii of this circle are $\overline{OA}, \overline{OC}, \overline{OB}$.

11. The three chords of this circle are $\overline{DA}, \overline{DC}, \overline{AC}$.

13. The two semicircles are $\widehat{ABC}, \widehat{ADC}$.

15. To find the diameter of a circle whose radius you know, double the radius.

17. a. The radius of this circle will be 1 inch.
    b. The diameter of this circle will be 2 inches.
    c. The circumference will be $2\pi r = 2\pi(1) = 2\pi$ inches.
    d. The area of the circle will be $\pi r^2 = \pi(1)^2 = \pi$ square inches

19. Squaring 6 should be the first operation.

Notation

21. The symbol $\widehat{AB}$ is read as **arc AB.**

23. The formula for the circumference of a circle is $C = \pi D$ or $C = 2\pi r$.

25. If C is the circumference of a circle and D is its diameter, then $\frac{C}{D} = \pi$.

27. 8π

Practice

29. $C = \pi D$
    $C = \pi(12)$
    $C \approx 37.70$ inches

31. $C = \pi D$

$D = \dfrac{C}{\pi}$

$D = \dfrac{36\pi}{\pi}$

$D = 36$ meters

33. The perimeter contributed by the two edges of the rectangular region is 16 feet. Consider the two semicircle regions together as one circle, $C = \pi D : C = \pi(3) : C \approx 9.42$. The total circumference is the sum of these or 25.42 feet.

35. The perimeter contributed by the rectangular region is $P = 2(8) + 6 = 22$ inches. The perimeter of the semicircle is $C = \frac{1}{2}\pi D : C = \frac{1}{2}\pi(6) : C \approx 9.42$. The total perimeter is the sum of these or 31.42 inches.

37. $A = \pi r^2 = \pi(3)^2 \approx 28.3$ square inches.

39. The area of the rectangular region is $A = 6(10) = 60$ square inches. Consider the two semicircles as a circle, the area of this circle is then $A = \pi r^2 = \pi(3)^2 \approx 28.3$ square inches. The total area is the sum of these or 88.3 square inches.

41. The area of the triangular region is $A = \frac{1}{2}bh = \frac{1}{2}(12)(12) = 72\ \text{cm}^2$. The area of the semicircle is $A = \frac{1}{2}\pi r^2 = \frac{1}{2}\pi\left(\frac{12}{2}\right)^2 \approx 56.5\ \text{cm}^2$. The total area is the sum of these or $128.5\ \text{cm}^2$.

43. The area of the rectangular region is $A = 10(4) = 40\ \text{in}^2$. The area of the circular region is $A = \pi r^2 = \pi(2)^2 \approx 12.6\ \text{in}^2$. The area of the shaded region is the difference of these or 27.4 in^2.

45. The area of the parallelogram is $A = 13(9) = 117\ \text{in}^2$. The area of the circle is $A = \pi r^2 = \pi(4)^2 \approx 50.3\ \text{in}^2$. The area of the shaded region is the difference of these or $66.7\ \text{in}^2$.

Applications

47. AREA OF ROUND LAKE

The area of the lake is approximately

$A = \pi r^2$ where $r = \frac{1}{2}(2)$, which is $A = \pi(1)^2 \approx 3.14\ \text{mi}^2$.

49. GIANT SEQUOIA

The diameter of this tree is $C = \pi D : D = \frac{C}{\pi} : D = \frac{102.6}{\pi} : D \approx 32.66$ feet.

51. JOGGING

One lap around the track is the circumference of the circular running surface or $C = \pi D : C = \pi(.25) : C \approx 0.7854$ miles. To run 10 miles on this track Joan must run $\frac{10}{0.7854} \approx 12.73$ laps.

53. BANDING THE EARTH

The circumference of the earth is $C = 2\pi r$, the new circumference will be $C + 10 = 2\pi r_1$, where r_1 is the new radius associated with the larger circumference. The difference between the two radii will yield the height of the band above the Earth's surface. $\frac{C+10}{2\pi} = r_1$ and

$$\frac{C}{2\pi} = r$$

$$r_1 - r = \frac{C+10}{2\pi} - \frac{C}{2\pi}$$

$$r_1 - r = \frac{10}{2\pi} \approx 1.59 \text{ feet}$$

55. ARCHERY

The area of the entire target is $A = \pi r^2 = \pi(2)^2 \approx 12.57$ ft^2. The area of the bull's eye is $A = \pi r^2 = \pi(0.5)^2 \approx 0.79$ ft^2. The bull's eye constitutes approximately $\frac{0.79}{12.57} \approx 0.0628$ or 6.28%.

Writing

57. Answers will vary.
59. Answers will vary.
61. Answers will vary.

Review

63. $\frac{9}{10} = 0.90$ or 90%

65. Convert $1.29 to cents, 129 cents. $\frac{129}{24} = 5.375$ cents per ounce.

67. A pentagon has five sides.

Section 9.7 Surface Area and Volume

Vocabulary

1. The space contained within a geometric solid is called its **volume**.
3. A **cube** is a rectangular solid with all sides of equal length.
5. The **surface** area of a rectangular solid is the sum of the areas of its faces.
7. A **cylinder** is a hollow figure like a drinking straw.
9. A **cone** looks like a witch's pointed hat.

Concepts

11. The volume of a rectangular solid is found using $V = lwh$.

13. The volume of a sphere is found using $V = \frac{4}{3}\pi r^3$.

15. The volume of a cone is found using $V = \frac{1}{3}\pi r^2 h$.

17. The surface area of a rectangular solid is found using $SA = 2lw + 2lh + 2hw$.

19. There are 27 ft^3 in a cubic yard. One cubic yard is $3 \times 3 \times 3 = 27$ ft^3.

21. There are 1,000 dm^3 in a cubic meter.

23. a. Volume should be applied for the size of a room to be air-conditioned.
    b. Area should be applied for the amount of land in a national park.
    c. Volume should be applied for the amount of space in a freezer.
    d. Surface area should be applied for the amount of cardboard in a shoebox.
    e. Perimeter should be applied for the distance around a checkerboard.
    f. Surface area should be applied for the amount of material to make a basketball.

25. a. The volume of this shape is $V = 3 \times 4 \times 6 = 72$ in^3.
    b. The area of the front of this shape is $A = 3 \times 6 = 18$ in^2.
    c. The area of the base of this shape is $A = 4 \times 6 = 24$ in^2.

Notation

27. The notation 1 in^3 is read as **1 cubic inch**.

Practice

29. $V = 3(4)(5) = 60 \text{ cm}^3$.

31. $V = \frac{1}{2}(3)(4)(8)$

    $V = 48 \text{ m}^3$

33. $V = \frac{4}{3}\pi(9^3)$

    $V \approx 3{,}053.63 \text{ in}^3$

35. $V = \pi(6^2)(12)$

    $V \approx 1{,}357.17 \text{ m}^3$

37. $V = \frac{1}{3}\pi(5^2)(12)$

    $V \approx 314.16 \text{ cm}^3$

39. $V = \frac{1}{3}(10^2)(12)$

    $V = 400 \text{ m}^3$

41. $SA = 2(4)(5) + 2(3)(4) + 2(3)(5)$

    $SA = 40 + 24 + 30$

    $SA = 94 \text{ cm}^2$

43. $SA = 4\pi(10^2)$

    $SA \approx 1{,}256.64 \text{ in}^2$

45. Consider this as a cube with a square pyramid. The volume of the cube is $V = 8^3 = 512 \text{ cm}^3$. The volume of the square pyramid is $V = \frac{1}{3}(8^2)(3) = 64 \text{ cm}^3$. The total volume is $512 + 64 = 576 \text{ cm}^3$.

47. Consider this as two identical cones. The volume of one of the cones is $V = \frac{1}{3}\pi(4^2)(10) \approx 167.55 \text{ in}^3$. Doubling this amount yields 335.103 in^3.

Applications

49. VOLUME OF A SUGAR CUBE

$$V = \left(\frac{1}{2}\right)^3 = \frac{1}{8} \text{ in}^3$$

51. WATER HEATER

Over 200 gallons of hot water from

$V = 27(17)(8) = 3,672 \text{ in}^3$. Converting to feet, $\frac{3,672}{12^3} = 2.125$ cubic feet of space.

53. VOLUME OF AN OIL TANK

$$V = \pi(3^2)(7) \approx 197.92 \text{ ft}^3$$

55. HOT-AIR BALLOON

$$V = \frac{4}{3}\pi(20^3) \approx 33,510.32 \text{ ft}^3$$

57. ENGINE

The compression ratio is $\frac{30.4}{3.8} = 8$, so the ratio is 8 : 1.

Writing

59. Answers may vary.

61. Answers may vary.

Review

63. $-5(5-2)^2 + 3 = -5(3)^2 + 3 = -45 + 3 = -42$

65. $\frac{x+7}{-4} = \frac{1}{4}$

$$x + 7 = -1$$

$$x = -8$$

67. $\frac{3}{15} = \frac{1}{5}$

69. 2.4 meters is 2,400 millimeters.

Chapter 9 Key Concept

1. $d = rt$

3. $P = 2l + 2w$

5. $A = \frac{1}{2}(700)(600) = 210,000 \text{ ft}^2$

7. $retail = 45.50 + 35 = 80.50$

9. $d = 16(3^2) = 144$ feet

11.

Type	Principal	Rate	Time	Interest
Savings	\$5,000	5%	3 yr	$I = 5000(0.05)(3) = \$750$
Passbook	2,250	2%	1	$I = 2250(0.02)(1) = \$45$
Trust Fund	10,000	6.25%	10	$I = 10000(0.0625)(10) = \$6,250$

Chapter 9 Review

Section 9.1 Some Basic Definitions

1. The points are C and D, the line is CD and the plane is GHI.

3. $\angle ABC, \angle CBA, \angle B, \angle 1$

5. Acute angles: $\angle 1, \angle 2$; Right angles $\angle ABD, \angle CBD$; Obtuse angle $\angle CBE$; Straight angle $\angle ABC$.

7. $x + 35 = 50$

   $x = 15^0$

9. a. $m\angle 1 = 65^0$

   b. $m\angle 2 = 180 - 65 = 115^0$

11. The supplement of a 140^0 angle is $180 - 40 = 40^0$.

Section 9.2 Parallel and Perpendicular Lines

13. Part *a* represents parallel lines.

15. The corresponding angles are $\angle 1$ and $\angle 5$, $\angle 4$ and $\angle 8$, $\angle 2$ and $\angle 6$, and $\angle 3$ and $\angle 7$.

17. $m\angle 1 = 70^0; m\angle 2 = 110^0; m\angle 3 = 70^0; m\angle 4 = 110^0; m\angle 5 = 70^0;$

    $m\angle 6 = 110^0; m\angle 7 = 70^0$

19. $x + 10 = 2x - 30$

    $40 = x$

Section 9.3 Polygons

21. a. This is an octagon.
    b. This is a pentagon.
    c. This is a triangle.
    d. This is a hexagon.
    e. This is a quadrilateral.

23. a. This is an isosceles triangle.
    b. This is a scalene triangle.
    c. This is an equilateral triangle.
    d. This is a right triangle.

25. a. $x + 70 + 20 = 180$

$$x = 90$$

b. $70 + 60 + x = 180$

$$x = 50$$

27. If one of the base angle is 60^0 then this is an equilateral triangle.

29. a. $m\overline{BD} = 15$ cm

b. $m\angle 1 = 40^0$

c. $m\angle 2 = 180 - 40 - 40 = 100^0$

31. a. $m\angle B = 65^0$

b. $m\angle C = 180 - 65 = 115^0$

Section 9.4 Properties of Triangles

33. $\angle A$ corresponds to $\angle D$
$\angle B$ corresponds to $\angle E$
$\angle C$ corresponds to $\angle F$
$\overline{AC}$ corresponds to $\overline{DF}$
$\overline{AB}$ corresponds to $\overline{DE}$
$\overline{BC}$ corresponds to $\overline{EF}$

35. a. These are similar triangles.
b. These are similar triangles.

37. a. $a^2 + b^2 = c^2$

$$5^2 + 12^2 = c^2$$

$$c^2 = 169$$

$$c = 13$$

b. $a^2 + b^2 = c^2$

$$8^2 + b^2 = 17^2$$

$$b^2 = 17^2 - 8^2$$

$$b^2 = 225$$

$$b = 15$$

Section 9.5 Perimeters and Areas of Polygons

39. $P = 4(18) = 72$ inches

41. a. $P = 8 + 4 + 4 + 8 + 6 = 30$ m
    b. $P = 10 + 4 + 4 + 4 + 6 + 8 = 36$ m

43. There are nine square feet in one square yard since 3 ft = 1 yd.

Section 9.6 Circles

45. a. The chords are $\overline{CD}$ and $\overline{AB}$.
    b. The diameter is $\overline{AB}$.
    c. The radii are $\overline{OA}$, $\overline{OC}$, $\overline{OD}$, and $\overline{OB}$.
    d. The center is O.

47. $P = 2(10) + 8\pi$
    $P \approx 45.1$ cm

49. $A = 10(8) + \pi(4^2)$
    $A \approx 130.3 \text{ cm}^2$

Section 9.7 Surface Area and Volume

51. There are $12^3 = 1,728 \text{ in}^3$ in one cubic foot.

53. a. $SA = 2(4.4)(3.1) + 2(2.3)(4.4) + 2(3.1)(2.3)$
       $SA = 61.78 \text{ ft}^2$

    b. $SA = 4\pi(5^2)$
       $SA \approx 314.2 \text{ in}^2$

Chapter 9 Test

1. $m(\overline{AB}) = 4$ units

3. This is true.

5. This is false.

7. $x + 17 = 67$

$$x = 50^0$$

9. $3y + 4 = 5y - 20$

$$24 = 2y$$
$$12 = y$$

11.The complement of 67^0 is $90 - 67 = 23^0$.

13. $m\angle 1 = 70^0$

15. $m\angle 3 = 70^0$

17.

Polygon	*Sides*
Triangle	3
Quadrilateral	4
Hexagon	6
Pentagon	5
Octagon	8

19. Since this is an isosceles triangle $m\angle A = 57^0$.

21. $x + 65 + 85 = 180$

$$x + 150 = 180$$
$$x = 30^0$$

23. $m(\overline{AB}) = m(\overline{DC})$

$$m(\overline{AD}) = m(\overline{BC})$$
$$m(\overline{AC}) = m(\overline{BD})$$

25. $m(\overline{DE}) = m(\overline{AB}) = 8$ in

27. $\frac{x}{9} = \frac{4}{6}$

$$x = \frac{36}{6}$$
$$x = 6$$

29. The straight line distance from third to first is $90^2 + 90^2 = d^2$

$$16{,}200 = d^2$$

$$127.28 \text{ ft} = d$$

31. $A = \frac{1}{2}(6)(12.2 + 15.7)$

$A = 83.7 \text{ ft}^2$

33. $A = \pi(3^2)$

$A \approx 28.27 \text{ ft}^2$

35. $V = \frac{4}{3}\pi(4^3)$

$V \approx 268.08 \text{ m}^3$

37. Answers may vary.

Chapters 1 – 9 Cumulative Review Exercises

1. AMUSEMENT PARKS

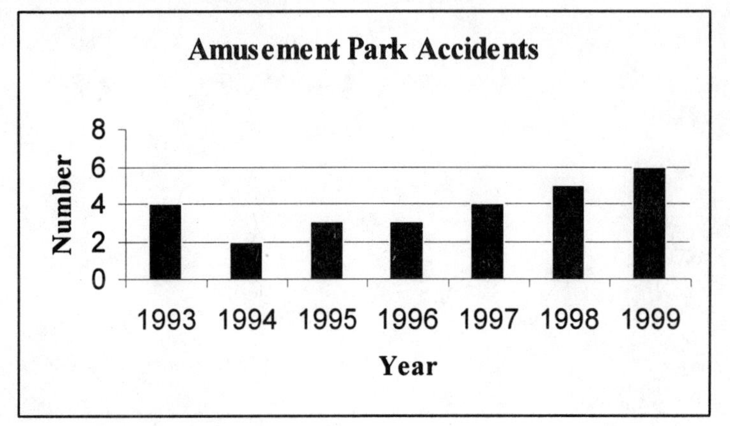

3. $35{,}021 - 23{,}999 = 11{,}022$

5. $2{,}110{,}000$

7. $F_{24} = \{1, 2, 3, 4, 6, 8, 12, 24\}$

9. $$\begin{aligned} -10(-2) - 2^3 + 1 &= 20 - 2^3 + 1 \\ &= 20 - 8 + 1 \\ &= 13 \end{aligned}$$

11. $$\begin{aligned} |-6 - (-3)| &= |-6 + 3| \\ &= |-3| \\ &= 3 \end{aligned}$$

13. An equation contains an equal sign and an expression does not.

15. $$\begin{aligned} 3(p + 15) + 4(11 - p) &= 0 \\ 3p + 45 + 44 - 4p &= 0 \\ -p + 89 &= 0 \\ 89 &= p \end{aligned}$$

17. $$\begin{aligned} -x + 2 &= 13 \\ -x &= 11 \\ x &= -11 \end{aligned}$$

19. SNAILS

In inches per minute the rate was $\frac{13}{2} = 6.5$ in/min .

21. 5 less than a number translates to $x-5$.

23. $\frac{35a^2}{28a}=\frac{5a}{4}$

25. $\frac{x}{4}-\frac{3}{5}=\frac{5x}{20}-\frac{12}{20}=\frac{5x-12}{20}$

27. $-\frac{6}{25}\left(2\frac{7}{24}\right)=-\frac{6}{25}\left(\frac{55}{24}\right)$

$=-\frac{\cancel{6}}{\cancel{5}(5)}\left(\frac{\cancel{5}(11)}{\cancel{6}(4)}\right)$

$=-\frac{11}{20}$

29. VETERINARY MEDECINE

A single dose is $\frac{\frac{3}{4}}{8}=\frac{3}{4}\bullet\frac{1}{8}=\frac{3}{32}$ ounce.

31. $\frac{2}{3}q-1=-6$

$\frac{2}{3}q=-5$

$q=\frac{3}{2}(-5)$

$q=-\frac{15}{2}$

33. $\frac{7-\frac{2}{3}}{4\frac{5}{6}}=\frac{6\frac{1}{3}}{4\frac{5}{6}}=\frac{\frac{19}{3}}{\frac{29}{6}}=\frac{19}{3}\bullet\frac{6}{29}=\frac{38}{29}$

35. GLOBAL WARMING

a. The greatest rise in temperature was recorded in 1991. The rise was about 0.3^0.

b. The greatest decline in temperature was recorded in 1993. The decline was about -0.5^0.

37. $\pi \approx 3.1416$

39. $3.4+106.78+35+0.008=145.188$

41. $(89.9708)(1,000)=89,970.8$

43. $-8.8+(-7.3-9.5)=-8.8+(-16.8)=-25.6$

45. $\frac{2}{15} = 0.1\bar{3}$

47. DECORATIONS

Let b represent the number of balloons.

$$\begin{aligned} 20 &= 15.15 + 0.05b \\ 4.85 &= 0.05b \\ 97 &= b \end{aligned}$$

She can buy 97 balloons.

49. $2\sqrt{121} - 3\sqrt{64} = 2(11) - 3(8) = 22 - 24 = -2$

51. TABLE TENNIS

The mean is $\frac{0.85 + 0.85 + 0.87 + 0.86 + 0.88 + 0.84 + 0.88 + 0.85}{8} = 0.86$ ounce.

The median is $\frac{0.85 + 0.86}{2} = 0.855$ ounce and the mode is 0.85 ounce.

53. The coordinates of the origin are (0, 0).

55. $y = -2x$

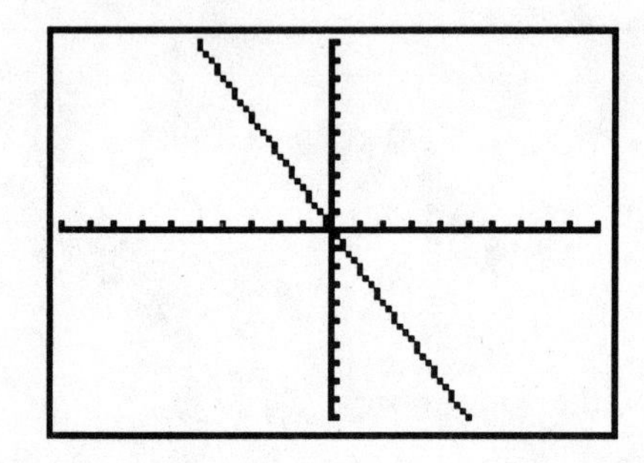

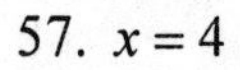

57. $x = 4$

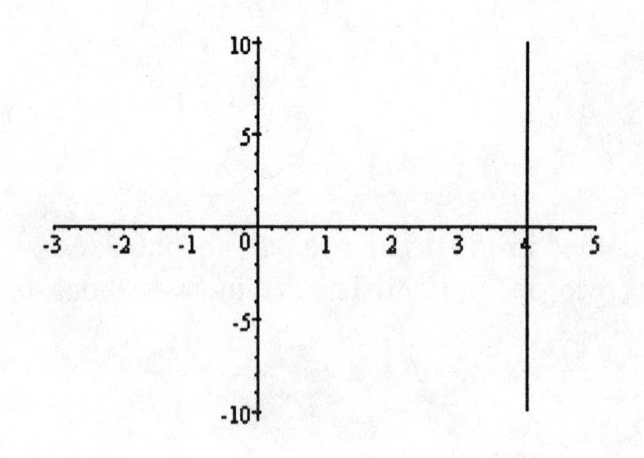

59. $-3(-2)^2 - 2(-2) = -3(4) + 4$
$= -12 + 4$
$= -8$

61. $s^4 s^5 = s^9$

63. $-3h^9(-5h) = 15h^{10}$ 64.

65. $(y^5)^2 (y^4)^3 = y^{10} y^{12} = y^{22}$

67. This is trinomial of degree two.

69. $-3p(2p^2 + 3p - 4) = -3p(2p^2) - 3p(3p) - 3p(-4)$
$= -6p^3 - 9p^2 + 12p$

71. $(2y-7)^2 = 2y(2y) + 2y(-7) + 2y(-7) + (-7)^2$
$= 4y^2 - 28y + 49$

73. $0.15(450) = 67.5$

75.

Percent	Decimal	Fraction
57%	0.57	$\frac{57}{100}$
0.1%	0.001	$\frac{1}{1,000}$
$33\frac{1}{3}\%$	$0.\bar{3}$	$\frac{1}{3}$

77. SHOPPING

The regular price is $(1 - 0.27)p = 54.75$
$0.73p = 54.75$
$p = \$75$

79. COLLECTIBLES

The percent increase was $125x = 750$
$x = 6$
$x = 600\%$

81. SAVING FOR RETIREMENT

$$A = 5,000\left(1 + \frac{0.08}{12}\right)^{50(12)} = \$269,390.92$$

83. $\frac{13}{52}=\frac{1}{4}$

85. $$\frac{5-x}{14}=\frac{13}{28}$$
$$28(5-x)=14(13)$$
$$5-x=-\frac{14(13)}{28}$$
$$x=-\frac{13}{2}+5$$
$$x=-\frac{3}{2}$$

87. SCALE DRAWING
There are 25 quarter inch squares on the length of the drawing therefore the house is $25(3)=75$ feet .

89. 15 yards $=15(3)(12)=540$ inches

91. 30 gallons $=30(14)=120$ quarts

93. 738 minutes $=\frac{738}{60}=12.3$ hours

95. 500 milliliters $=\frac{500}{1000}=0.5$ liter

97. 75^0 C $=\frac{5}{9}(75)+32=167^0$ F

99. TENNIS
$57 \text{ g} = 5{,}700 \text{ cg}$
$58 \text{ g} = 5{,}800 \text{ cg}$

101. COOKING
A 10-pound ham weighs about $\frac{10}{2.2}=4.5$ kilograms.

103. An acute angle is between zero and ninety degrees.

105. The complement of an angle of 75 degrees is $90-75=15$ degrees.

107. $m\angle 2=130^0$

109. $m\angle 4=m\angle 3=50^0$

111. $m\angle C=30^0$ since this is an isosceles triangle.

113. $m\angle 3 = 180 - 75 = 105^0$

115. The sum of the interior angles of a pentagon are $(5-2)180 = 540^0$.

117. $P = 2(9) + 2(12) = 42$ m

$A = 9(12) = 108 \text{ m}^2$

119. $A = \frac{1}{2}(7)(12+14)$

$A = \frac{1}{2}(7)(26)$

$A = 7(13)$

$A = 91 \text{ in}^2$

121. First find the area of the rectangle, then subtract out the area of the semicircles.

$A = 20.2(19.2) - \pi\left(\frac{19.2}{2}\right)^2$

$A \approx 98.31 \text{ yd}^2$

123. $V = \frac{4}{3}\pi(5)^3$

$V \approx 523.60 \text{ m}^3$

125. $V = \pi(0.25)^2(20)$

$V \approx 15.71 \text{ft}^3$

Appendix 1 Inductive and Deductive Reasoning

Vocabulary

1. **Inductive** reasoning draws general conclusions from specific observations.

Concepts

3. The pattern is circular.
5. The pattern is alternating.
7. The pattern is alternating.

9. ROOM SCHEDULING
   The room is available on Wednesday at 10 a.m.

Practice

11. $1,5,9,13,\underline{17},\ldots$
13. $-3,-5,-8,-12,\underline{-17},\ldots$
15. $-7,9,-6,8,-5,7,-4,\underline{6},\ldots$
17. $9,5,7,3,5,1,\underline{3},\ldots$
19. $-2,-3,-5,-6,-8,-9,\underline{-11},\ldots$
21. $6,8,9,7,9,10,8,10,11,\underline{9},\ldots$
23.

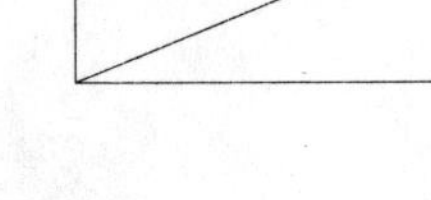

25.

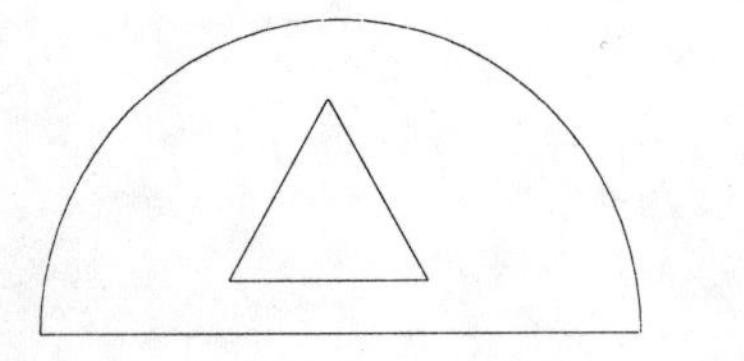

27. $A,c,E,g,\underline{I},\ldots$
29. $d,h,g,k,j,n,\underline{m},\ldots$
31. Maria must be the teacher since we know the baker is Luis and that Maria is unmarried.
33. Think of the four vehicles in spaces one through four, left to right. The Buick is in space 1, for the Ford to be between the Dodge and Mercedes, the Mercedes must be in space 4, the right end.

35. From top to bottom the flag colors are green, blue, yellow, then red.

Applications

37. JURY DUTY

There are 18,935 respondents who have served on neither a criminal court nor a civil jury, see the Venn diagram.

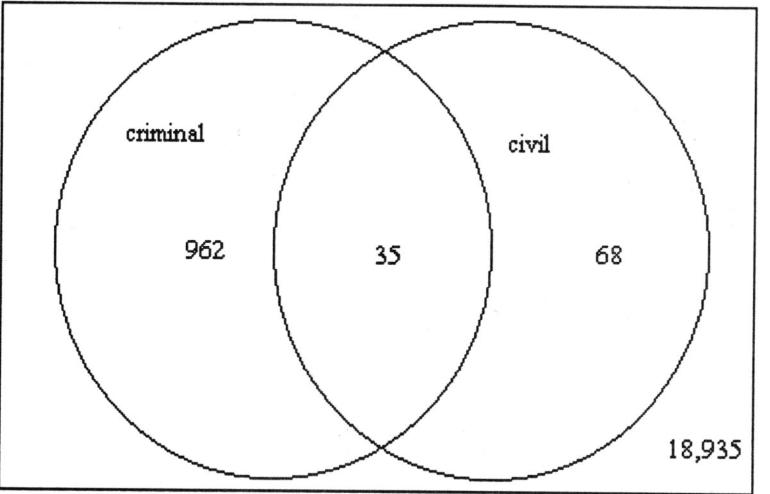

39. THE SOLAR SYSTEM

There are zero planets that are neither rocky nor have moons.

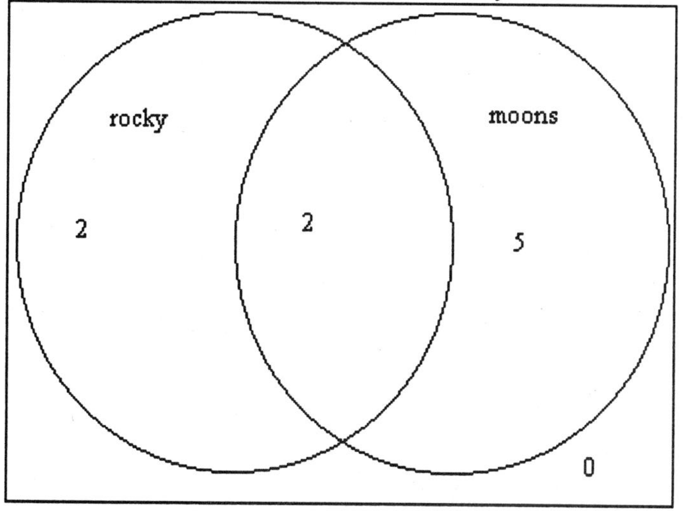

Writing

41. Answers may vary.
43. Answers may vary.